Origin and Evolution of Metazoan Cell Types

Evolutionary Cell Biology

Series Editors

Brian K. Hall – *Dalhousie University, Halifax, Nova Scotia, Canada*

Sally A. Moody – *George Washington University, Washington DC, USA*

Editorial Board

Michael Hadfield – *University of Hawaii, Honolulu, USA*

Kim Cooper – *University of California, San Diego, USA*

Mark Martindale – *University of Florida, Gainesville, USA*

David M. Gardiner – *University of California, Irvine, USA*

Shigeru Kuratani – *Kobe University, Japan*

Nori Satoh – *Okinawa Institute of Science and Technology, Japan*

Sally Leys – *University of Alberta, Canada*

Science publisher

Charles R. Crumly – *CRC Press/Taylor & Francis Group*

Published Titles

Cells in Evolutionary Biology: *Translating Genotypes into Phenotypes – Past, Present, Future*
Edited by Brian K. Hall and Sally Moody

Deferred Development: Setting Aside Cells for Future Use in Development in Evolution
Edited by Cory Douglas Bishop and Brian K. Hall

Cellular Processes in Segmentation
Edited by Ariel Chipman

Cellular Dialogues in the Holobiont
Edited by Thomas Bosch and Michael G. Hadfield

Evolving Neural Crest Cells
Edited by Daniel Meulemans Medeiros, Brian Frank Eames, Igor Adameyko

For more information about this series, please visit: www.crcpress.com/Evolutionary-Cell-Biology/book-series/CRCEVOCELBIO

Origin and Evolution of Metazoan Cell Types

Edited by
Sally Leys and Andreas Hejnol

CRC Press is an imprint of the
Taylor & Francis Group, an **informa** business

First edition published 2021
by CRC Press
6000 Broken Sound Parkway NW, Suite 300, Boca Raton, FL 33487-2742
and by CRC Press
2 Park Square, Milton Park, Abingdon, Oxon, OX14 4RN

© 2021 Taylor & Francis Group, LLC

CRC Press is an imprint of Taylor & Francis Group, LLC

Reasonable efforts have been made to publish reliable data and information, but the author and publisher cannot assume responsibility for the validity of all materials or the consequences of their use. The authors and publishers have attempted to trace the copyright holders of all material reproduced in this publication and apologize to copyright holders if permission to publish in this form has not been obtained. If any copyright material has not been acknowledged, please write and let us know so we may rectify in any future reprint.

Except as permitted under U.S. Copyright Law, no part of this book may be reprinted, reproduced, transmitted, or utilized in any form by any electronic, mechanical, or other means, now known or hereafter invented, including photocopying, microfilming, and recording, or in any information storage or retrieval system, without written permission from the publishers.

For permission to photocopy or use material electronically from this work, access www.copyright.com or contact the Copyright Clearance Center, Inc. (CCC), 222 Rosewood Drive, Danvers, MA 01923, 978-750-8400. For works that are not available on CCC, please contact mpkbookspermissions@tandf.co.uk

Trademark notice: Product or corporate names may be trademarks or registered trademarks and are used only for identification and explanation without intent to infringe.

ISBN: 9781138032699 (hbk)
ISBN: 9780367766085 (pbk)
ISBN: 9781315388229 (ebk)

Typeset in Times
by Deanta Global Publishing Services, Chennai, India

Contents

Series Preface vii
Preface ix
Acknowledgements xiii
Editor Biographies xv
List of Contributors xvii

Chapter 1 What Is a Cell Type? 1
  Alessandro Minelli

Chapter 2 The Protistan Origins of Animal Cell Differentiation 13
  Sebastián R. Najle and Iñaki Ruiz-Trillo

Chapter 3 Convergent Evolution of Animal-Like Organelles across the Tree of Eukaryotes 27
  Greg S. Gavelis, Gillian H. Gile and Brian S. Leander

Chapter 4 Evolution of the Animal Germline: Insights from Animal Lineages with Remarkable Regenerating Capabilities 47
  Ana Riesgo and Jordi Solana

Chapter 5 Origin and Evolution of Epithelial Cell Types 75
  Emmanuelle Renard, André Le Bivic, and Carole Borchiellini

Chapter 6 Evolution of the Sensory/Neural Cell Types 101
  Sally P. Leys, Jasmine L. Mah, Emma K.J. Esposito

Chapter 7 Cell Types, Morphology, and Evolution of Animal Excretory Organs 129
  Carmen Andrikou, Ludwik Gąsiorowski, and Andreas Hejnol

Index 165

Series Preface

In recent decades, evolutionary principles have been integrated into biological disciplines such as developmental biology, ecology and genetics. As a result, major new fields emerged, chief among which are Evolutionary Developmental Biology (or Evo-Devo) and Ecological Developmental Biology (or Eco-Devo). Inspired by the integration of knowledge of change over single life spans (ontogenetic history) and change over evolutionary time (phylogenetic history), evo-devo produced a unification of developmental and evolutionary biology that generated unanticipated synergies: Molecular biologists employ computational and conceptual tools generated by developmental biologists and by systematists, while evolutionary biologists use detailed analysis of molecules in their studies. These integrations have shifted paradigms and enabled us to answer questions once thought intractable.

Major highlights in the development of modern Evo-Devo are a comparison of the evolutionary behavior of cells, evidenced in Stephen J. Gould's 1979 proposal of changes in the timing of the activity of cells during development—heterochrony—as a major force in evolutionary change, and numerous studies demonstrating how conserved gene families across numerous cell types "explain" development and evolution. Advances in technology and in instrumentation now allow cell biologists to make ever more detailed observations of the structure of cells and the processes by which cells arise, divide, differentiate and die. In recent years, cell biologists have increasingly asked questions whose answers require insights from evolutionary history. As just one example: How many cell types are there and how are they related? Given this conceptual basis, cell biology—a rich field in biology with history going back centuries—is poised to be reintegrated with evolution to provide a means of organizing and explaining diverse empirical observations and testing fundamental hypotheses about the cellular basis of life. Integrating evolutionary and cellular biology has the potential to generate new theories of cellular function and to create a new field, *"Evolutionary Cell Biology."*

Mechanistically, cells provide the link between the genotype and the phenotype, both during development and in evolution. Hence the proposal for a series of books under the general theme of *"Evolutionary Cell Biology: Translating Genotypes into Phenotypes"*, to document, demonstrate and establish the central role played by cellular mechanisms in in the evolution of all forms of life.

Brian K. Hall and Sally A. Moody

Preface

Animals intrigue us with their different lifestyles, the many ways they are adapted to their environments, and the fascinating means by which they survive and reproduce. Animals tend to have complex behavior, and the ways in which this behavior is carried out makes them different from other multicellular branches such as plants and fungi. This volume focuses on the evolution of metazoan cell types that have given rise to this diversity of animal body plans, lifestyles, and complex behaviours.

Cells are the units of life, and with the evolution of multicellularity, cells were able to undergo a tremendous diversification and functional specialization. These novel cellular features are made possible by the evolution of distinct cell types that are highly specialized, for example, for protection from the environment, uptake of nutrients, transport, sensing and transmitting information, and reproduction, and by their interactions. Expansion into ecological niches, competition, and predation among early animals provided the framework for continuous evolutionary change, transforming simple epithelia into transport tubes, and specialization of cells as blood for transporting nutrients and gases, sequestering the germ lineage in cells and organs, and distinguishing cells for paracrine, endocrine, and neuronal signalling.

It has been argued (Valentine et al., 1994) that the number of cell types is an effective measure of organismal complexity with vertebrates having more types and more complexity than invertebrates and non-bilaterians. Such hierarchical views of complexity have dominated the evo-devo discussion on the evolution of metazoan cell types and body plans (Scholtz, 2004; Dunn et al., 2015). And while there is good evidence from a broad range of studies that cell types are conserved within some lineages (Erkenbrack and Thompson, 2019), it turns out that it is difficult to detect homologies between cell types across longer evolutionary distances (Sebé-Pedrós et al., 2018a; Sebé-Pedrós et al., 2018b). Convergence, not very long ago considered unlikely, is now widely accepted to have happened at all biological levels (cells, tissues, body plans).

Cellular evolution shares parallels across the different multicellular groups (plants, fungi, animals), for example, mechanisms of adhesion in epithelia of slime molds (*Dictyostelium*) and placozoans (Dickinson et al., 2012; Weis et al., 2013; Smith and Reese, 2016). Although in many cases multiple origins are inferred for cell types with the same function, it seems likely that some already specialized cell types in the earliest metazoans do share a common ancestry (are homologous). Discriminating between homology and homoplasy of cell types has always been problematic due to the lack of a resolved phylogeny of organisms, but new tools have been developed in recent years that go beyond ultrastructural comparisons and cellular physiology.

While ultrastructural similarity and cellular function were previously used to determine cell types (Valentine et al., 1994), recent effort has been in sequencing, and most recently in single cell RNAseq methods. Although these data bring a new perspective to the definition of a cell type, they also add some blur to the picture and do not in fact increase our understanding of the basic categories used. Surprisingly, estimates of the minimum number of cell types in different organisms (not including neuronal subtypes) in the human body has not changed from Valentine's (1994) estimates even with scRNAseq techniques (Trapnell, 2015; Cao et al., 2017). Instead, with the introduction of sequencing approaches to investigations of cell complements of organisms, new definitions, such as cell states and sub-cell-types have had to be introduced to provide a better semantic framework that describes the biological observations. This has led to the co-existence of different perspectives what a cell type actually is (e.g., Mak, 2017).

Comparative developmental data using cell lineage tracers, gene expression, single cell genomics/transcriptomics, and gene knockout approaches can provide support for shared gene networks that designate specific cell types as similar to each other. A less studied data set includes physiological function determined by complement of ion channels, receptors, paracrine molecules, and even cell behavior. These tools however, are also influencing the picture of what we understand and identify as a cell type and could provide a level of novel definition.

What is special about the evolution of cell types as a unit is that each cell in the body carries the genomic information that contains all modules used by other cell types. Since this information is also inherited by the next generation via the germ line, we can expect the independent and recurrent emergence of similar cell types in many lineages. To discriminate between homology and convergence on the level of cell types will be one of the greatest challenges in the future, because of the easiness of the horizontal transfer of cellular submodules by activation of cis-regulatory elements. This is similar to what is observed in the vast exchange of genetic material between species in prokaryotes which still poses the major problem in resolving prokaryote phylogeny, the aptly named "ring of life" (Rivera and Lake, 2004).

Why do we need to know what the origin of a particular cell type is? The evolution of animal diversity is likely strongly affected by the origin of novel cell and tissue types and their interactions with each other. Understanding the evolution of cell types will shed light on the evolution of novel structures and in turn highlight how animals diversified. Several cell types may also have been lost as animals simplified, and a better grasp of the evolutionary history of cell types and how new cell types evolve will allow a better understanding that can discriminate between convergence and homology.

The target audience for this volume is a diverse group that functions at the graduate level and higher, and which crosses all disciplines—physiology, cell, chemical, and molecular biology—and who will be seeking a new approach to understanding the evolution of animal diversity.

In Chapter 1, we start with a stimulating analysis of "What a cell type is." Here, Alessandro Minelli from the University of Padova, Italy, explores the history, philosophy, and semantics of classification of cells by molecules and morphology. He also considers how cell types change with space and time and concludes that we might need to be content with "some degree of taxonomic pluralism" in our understanding of cell type systematics.

The next two chapters examine the protistan origins of animal cell types. In Chapter 2, Najle and Ruiz-Truillo from the CSIC, University of Barcelona and the University of Pompeu Fabra, Spain, argue that unicellular holozoans have different cell states which are equal to cell types, only so far, these cell states are not known to exist at the same time in the same organism. In Chapter 3, Gavelis, Gile, and Leander from the University of British Columbia, Canada, highlight some particularly complex differentiated cell types in unicellular eukaryotes that function as sensory structures. They look at statocysts, eye-like structures, and harpoon type nematocyst-like structures in different protist groups and argue that homology plays little role in how these structures came about.

Cell types tend to have different phenotypes, whether morphological, molecular, or physiological. But during animal development, all cell types arise from the fusion of the two gametes, and gametes arise from germ cells. How did the germ cell type arise in metazoan evolution? In Chapter 4, Riesgo and Solana, from the Natural History Museum London and Oxford Brooks University, England, focus on early-branching phyla to look for commonalities of germ cell segregation in animals and argue that molecular mechanisms for regeneration hold a clue to how this cell type may have arisen.

One of the more recognizable cell types in animals is the epithelial cell, since it is what lines all tissues and gives sensory, glandular properties to many organs. Yet epithelia are so diverse across metazoans, and in Chapter 5, we learn from Renard, Le Bivic, and Borchiellini, from the University of Aix Marseille, France, that even agreement on what constitutes an epithelium is disputed.

Chapter 6 addresses another iconic metazoan cell type, the neuron. Here Leys, Mah, and Esposito, from the University of Alberta, Canada, examine what a neuron is and assess the morphological and molecular character of neurons as well as sensory cells in non-bilaterians. The problem of bias in interpreting new scRNAseq data is discussed.

In the last chapter, Chapter 7, Andrikou, Gąsiorowski, and Hejnol from the University of Bergen, Norway, examine the origin of nephridia with a careful review of the morphology and transcriptomic profiles of excretory cells in invertebrates. They argue that homology of organelles and even of some excretory cell types does not imply homology of the organs, and they caution against the use of gene combinations to detect cell types without supporting functional data.

LITERATURE CITED

Cao, J., Packer, J. S., Ramani, V., Cusanovich, D.A., Huynh, C., Daza, R., Qiu, X., Lee, C., Furlan, S. N., Steemers, F. J., Adey, A., Waterston, R. H., Trapnell, C. and Shendure, J. 2017. Comprehensive single-cell transcriptional profiling of a multicellular organism. *Science* 357:661–667. doi:10.1126/science.aam8940

Dickinson, D. J., Nelson, W. J. and Weis, W. I. 2012. An epithelial tissue in *Dictyostelium* challenges the traditional origin of metazoan multicellularity. *BioEssays* 34:833–840. doi:10.1002/bies.201100187

Dunn, C. W., Leys, S. P. and Haddock, S. H. D. 2015. The hidden biology of sponges and ctenophores. *Trends Ecol. Evol.* 30:282–291. doi:10.1016/j.tree.2015.03.003

Erkenbrack, E. M. and Thompson, J. R. 2019. Cell type phylogenetics informs the evolutionary origin of echinoderm larval skeletogenic cell identity. *Commun. Biol.* 2:160. doi:10.1038/s42003-019-0417-3

Mak, H. C. 2017. Clouds, continuums, and cells. *Cell Syst.* 4:251. doi:10.1016/j.cels.2017.03.008

Rivera, M. C. and Lake, J. A. 2004. The ring of life provides evidence for a genome fusion origin of eukaryotes. *Nature* 431:152–155. doi:10.1038/nature02848

Scholtz, G. 2004. Baupläne versus ground patterns, phyla versus monophyla: Aspects of patterns and processes in evolutionary developmental biology. In Scholtz, G. (Ed.), *Evolutionary Developmental Biology of Crustacea*. Balkema: Lisse: 3–16.

Sebé-Pedrós, A., Chomsky, E., Pang, K., Lara-Astiaso, D., Gaiti, F., Mukamel, Z., Amit, I., Hejnol, A., Degnan, B. M. and Tanay, A. 2018a. Early metazoan cell type diversity and the evolution of multicellular gene regulation. *Nat. Ecol. Evol.* 2:1176–1188. doi:10.1038/s41559-018-0575-6

Sebé-Pedrós, A., Saudemont, B., Chomsky, E., Plessier, F., Mailhé, M.-P., Renno, J., Loe-Mie, Y., Lifshitz, A., Mukamel, Z., Schmutz, S., Novault, S., Steinmetz, P. R. H., Spitz, F., Tanay, A. and Marlow, H. 2018b. Cnidarian cell type diversity and regulation revealed by whole-organism single-cell RNA-seq. *Cell* 173:1520–1534. e1520. doi:10.1016/j.cell.2018.05.019

Smith, C. L. and Reese, T. S. 2016. Adherens junctions modulate diffusion between epithelial cells in *Trichoplax adhaerens*. *Biol. Bull.* 231:216–224. doi:10.1086/691069

Trapnell, C. 2015. Defining cell types and states with single-cell genomics. *Genome Res.* 25:1491–1498.

Valentine, J. W., Collins, A. G. and Meyer, C. P. 1994. Morphological complexity increase in metazoans. *Paleobiology* 20:131–142. doi:10.1017/S0094837300012641

Weis, W. I., Nelson, W. J. and Dickinson, D. J. 2013. Evolution and cell physiology. 3. Using *Dictyostelium discoideum* to investigate mechanisms of epithelial polarity. *Am. J. Physiol. Cell Physiol.* 305:C1091–C1095. doi:10.1152/ajpcell.00233.2013

Acknowledgements

We are grateful to the many reviewers who helped improve this volume through comments on the different chapters. We also thank Brian Hall for proposing this challenging topic and the authors of each chapter, who accepted the challenge and in doing so provide us with unique insights into a range of metazoan cell types, their associations, and potential origins. Concepts developed here arose from work funded by an NSERC Discovery Grant (2016-05446 SPL) and are inspired by the Marie Sklodowska Curie Innovative Training Network "EvoCELL" (Grant Agreement No. 766053 to AH).

Editor Biographies

Andreas Hejnol is Professor and research group leader of "Comparative Developmental Biology" at the Department of Biological Sciences (BIO) in Bergen, Norway. After earning his Ph.D. in Comparative Zoology from the Free University Berlin, Germany in 2002, he worked as a postdoctoral fellow in the laboratory of Ralf Schnabel in Braunschweig and at the Kewalo Marine Laboratory in the lab of Mark Q. Martindale in Hawaii. He led a research group at the Sars Centre from 2009-2019. His research aims to understand the evolutionary origin and diversification of animal body plans, cell types, and organ systems. He is an ERC Consolidator Grant holder and received for his achievements in Evolutionary Developmental Biology and Comparative Zoology the prestigious Alexander O. Kovalevsky Medal from the St. Petersburg Society for Naturalists in 2018.

Sally P. Leys is Professor in the Department of Biological Sciences at the University of Alberta, in Edmonton, Canada. She earned her Ph.D. from the University of Victoria under George Mackie in 1996, for which she received the Canadian Society of Zoologists Cameron Award 1997. She held a Commander C Bellairs Postdoctoral Fellowship from McGill University for postdoctoral research in Barbados (1997) and then won an NSERC PDF which she took to the University Aix Marseille, France (1998) and later to the University of Queensland, Australia (1998-2000). She won an NSERC Women's University Research Award in 2000 and was Assistant Professor (Limited Term) at the University of Victoria, British Columbia. In 2002, she was awarded a Canada Research Chair Tier II at the University of Alberta in "Evolutionary and Developmental Biology." Her research interests broadly concern understanding the origin of multicellularity in metazoans and more specifically the cellular and molecular basis of coordination in non-bilaterian animals, sponges, ctenophores, placozoans, and cnidarians.

List of Contributors

Carmen Andrikou
Department of Biological Sciences
University of Bergen
Bergen, Norway

Carole Borchiellini
Aix Marseille Univ
CNRS, IRD
Avignon Université
Institut Méditerranéen de Biodiversité et d'Ecologie marine et continentale (IMBE)
Marseille, France

Emma K.J. Esposito
Department of Biological Sciences
University of Alberta
Edmonton, Canada

Ludwik Gąsiorowski
Department of Biological Sciences
University of Bergen
Bergen, Norway

Greg S. Gavelis
Departments of Botany and Zoology
Biodiversity Research Centre and Museum
University of British Columbia
Vancouver, British Columbia, Canada
and
School of Life Sciences
Arizona State University
Tempe, Arizona

Gillian H. Gile
School of Life Sciences
Arizona State University
Tempe, Arizona

Andreas Hejnol
Department of Biological Sciences
University of Bergen
Bergen, Norway

Brian S. Leander
Departments of Botany and Zoology
Biodiversity Research Centre and Museum
University of British Columbia
Vancouver, British Columbia, Canada

André Le Bivic
Aix Marseille Univ
CNRS
Institute of Developmental Biology of Marseille (IBDM)
Marseille, France

Sally P. Leys
Department of Biological Sciences
University of Alberta
Edmonton, Canada

Jasmine L. Mah
Department of Ecology and Evolutionary Biology
Yale University, Connecticut

Alessandro Minelli
University of Padova
Department of Biology
Padova, Italy

Sebastián R. Najle
Institut de Biología Evolutiva (CSIC-
  Universitat Pompeu Fabra)
Barcelona, Catalonia, Spain
and
Centre for Genomic Regulation (CRG)
Barcelona Institute of Science and
  Technology (BIST)
Barcelona, Spain

Ana Riesgo
Department of Life Sciences
The Natural History Museum of
  London
London, UK

Emmanuelle Renard
Aix Marseille Univ
CNRS
Institute of Developmental Biology of
  Marseille (IBDM)
Marseille, France

Iñaki Ruiz-Trillo
Institut de Biología Evolutiva (CSIC-
  Universitat Pompeu Fabra)
Barcelona, Catalonia, Spain
and
Departament de Genètica,
  Microbiologia i Estadística
Universitat de Barcelona
Barcelona, Catalonia, Spain
and
ICREA
Barcelona, Catalonia, Spain

Jordi Solana
Department of Biological and Medical
  Sciences
Faculty of Health and Life Sciences
Oxford Brookes University
Oxford, UK

1 What Is a Cell Type?

Alessandro Minelli

CONTENTS

1.1 Changing Criteria for the Recognition of Cell Types 1
1.2 Steady Renewal vs. Terminal Differentiation 2
1.3 Homology Problems 4
1.4 Cell Type as a Nomadic Concept 5
1.5 A Tentative Conclusion: Cell Type as Unit of Representation 6
References 8

1.1 CHANGING CRITERIA FOR THE RECOGNITION OF CELL TYPES

Since the beginning of systematic use of microscopic techniques in the study of animal tissues, there have been efforts to classify cells based on morphology and attempts to identify the functions associated with the different cell types. Specifically, since the mid-19th century, a classification of human cell types based on morphology and function has been developed because of its importance for human pathology and therefore for medicine—witness Rudolf Virchow's program of cellular pathology (*Cellularpathologie*; this is also the title of his 1858 book). However, a classification of cell types that is useful for humans does not necessarily apply beyond the animal groups closest to our species.

Eventually, despite the fact that morphological information is generally much more readily accessible than evidence about function, the latter has largely remained the first criterion for classification of cell types, at least at the coarsest level (neurons, muscular fibers, secretory cells etc.). We could therefore still say that cell types are populations of cells dedicated to different functional roles (Blainey, 2017; Wagner, 2019).

However, morphology and function are not always congruent. This is true even for morphologically unique cells such as choanocytes (collar cells), that are regarded by many authors (e. g. Richter and King, 2013; Arendt et al., 2016; Brunet and King, 2017) as one of the oldest cell types characteristic of the animal kingdom. In sponges, the movement of the flagellum of choanocytes favors phagocytosis of bacteria etc. by pushing solid particles suspended in water towards the cell; in other metazoans, however, cells morphologically identifiable as choanocytes have very different functional roles, e.g. sensory or excretory (Brunet and King, 2017; Sebé-Pedrós et al., 2018). Eventually, irrespective of the preference accorded in principle to this criterion, a classification of cell types based on function is often unattainable in practice (Lundberg and Uhlen, 2017; Sanes, 2017), so other criteria must be used.

For some years now, a molecular characterization of cells has been added to the traditional descriptors. Atlases of the molecular diversity of cell types have been obtained for some species. For example, the transcription patterns recognizable at the single-cell level allow us to recognize 38 major clusters that Cao et al. (2019) qualify as cellular types, with a total number of 655 subtypes, in mice.

According to Xia and Yanai (2019), a cell type is defined by the fraction of the genome it expresses, that is, by its transcripts, by the proteins or peptides that derive from the translation of the former, and by the network of relationships that bind the products of transcription; however, these authors eventually conclude that, operationally, every cell type is adequately defined by the transcription factors it expresses. For Arendt et al. (2019), in defining cell types, it is instead preferable to prioritize the mechanism that regulates transcription rather than the expressed effector genes. Borriello et al. (2020) define a cell type as a collection of attractors of a gene regulatory network (GRN), i.e. as the phenotypic expression of the steady states of the regulatory process, whereas cell behavior like apoptosis and proliferation is the phenotypic expression of the corresponding dynamics.

Shifting focus from morphology to the characterization of GRNs or transcriptomes, what changes is not only the set of tools used to explore cell diversity—from microscopes to sequencing apparatuses—but also the kind of taxonomy in which the observations are collected: instead of tissues characterized by cell types differing in form and function, we find transcriptionally coherent clusters or metacells (Sebé-Pedrós et al., 2018) within which the global transcription patterns are sufficiently homogeneous and sufficiently distinct from those characterizing the other clusters. This analysis applies equally well to sponges as it does to animals with tissue organization, as demonstrated by recent papers such as Sebé-Pedrós et al. (2018), where the cell types recognizable on a transcriptomic basis in the sponge *Amphimedon queenslandica*, the ctenophore *Mnemiopsis leidyi*, and the placozoan *Trichoplax adhaerens* are compared.

In *Mnemiopsis leidyi*, the patterns recognizable based on transcriptomes discloses a diversity of cell types, most of which cannot be associated with cell types distinguished by morphology and function (Sebé-Pedrós et al., 2018). Even in the case of organisms with a particularly simple organization, formed by a few thousand cells only, the diversity of cell types suggested by transcriptomes is different (higher) than the diversity suggested by morphology. In *Trichoplax adhaerens*, Sebé-Pedrós et al. (2018) recognize 11 cell types on a molecular basis, compared to the 7 types identified based on morphology (Smith et al., 2014; Schierwater and DeSalle, 2018).

1.2 STEADY RENEWAL VS. TERMINAL DIFFERENTIATION

Enthusiasm for a taxonomy of cell types based on molecular descriptors is moderated by the remark (e.g., Eberwine, 2017) that these descriptors (especially transcriptome, but also epigenome, proteome, or metabolome) only provide a static description of the cell. To some extent, to be sure, this is also true of morphology.

A way to save the notion of cell type and thus to fix a list of cell types recognized based either on morphology or on molecular signatures in a given animal species, even in front of the developmental transitions a single cell may undergo, is to strictly focus on a cell's terminal differentiated state, as far as this notion applies. It is easier, of course, to classify well-differentiated cells rather than those that are in a non-terminal phase of a differentiation process, for example, those that represent non-terminal phases of spermatogenesis (Xia and Yanai, 2019).

The notion of terminal differentiation presupposes an adultocentric scenario (Minelli, 2003), in which the history of the individual begins with an egg, fertilized or not, and proceeds towards the adult condition through successive stages of increasing morphological complexity and increasing diversity of cell types. However, this scenario does not apply universally to all metazoans. For example, it would be quite difficult to identify states of terminal differentiation in animals reproducing asexually by fragmentation (Wagner, 2019); the value of these models is also questionable in the case of tissues produced during regeneration. More disturbing, however, are animals such as the sea squirt *Botryllus*, in which a new individual can develop through either of two dramatically different ontogenetic paths: that is, either from the egg or from a bud (Manni and Burighel, 2006; Alié et al., 2020).

Indeed, some molecular models proposed to explain how the terminal differentiation of a cell would be achieved, i.e. the core gene regulatory network model of Graf and Enver (2009) and the terminal selector gene model of Hobert (2011), were largely based on selected cell types of vertebrates, neurons especially. General applicability has, however, been declared for the kernel model of Davidson and Erwin (2006) and the character identity network model proposed by Wagner (2007). These models identify a regulatory system that provides for a core gene regulatory network that operates under the influence of inductive signals.

The developmental history of a differentiated cell that we eventually classify as belonging to a specified cell type is very different in animals belonging to different phyla. The strict and thus predictable determination of cell types in *Caenorhabditis elegans* is obtained through a series of switches that correspond to as many mitotic divisions along a robust cell lineage, while the fixation of cell types in a vertebrate is not rigidly associated with either the cell's position in the cell lineage tree or with a series of irreversible switches associated with the cell divisions from which it is derived.

In *Amphimedon queenslandica*, the transition from choanocyte to archaeocyte does not require a mitosis (Sogabe et al., 2019); similar to trypanosomes, where transitions between different developmental stages are not separated by a cell division: for example, the transition from amastigote (without flagellum) to epimastigote (with flagellum attached to the cell body by a short membrane) is a process of (reversible) morphological differentiation (Matthews, 2005).

The pattern of cell types recognizable in an animal changes anyway along its ontogeny. Results of recent studies in which this aspect is studied at a molecular level are quite impressive. In *Amphimedon queenslandica*, only one cell type (archaeocytes) is recognizable both in the larva and the adult, compared to at least seven larval cell types (Sebé-Pedrós et al., 2018) and a higher number of adult cell types. This corresponds to large differences in the total number of genes expressed in the two stages. Furthermore, compared to 39.9% of the 22,567 genes expressed in the adult that are not expressed in the larva, there is also a 4.8% of the 14,426 genes expressed in the larva that are no longer expressed in the adult.

Of particular interest is the behavior of the choanocytes of this sponge, which, in spite of the easily recognizable morphological and functional characteristics that make them an almost ideal example of cellular type, actually represent a metastable condition; they easily change into cells substantially comparable to stem cells, ready to differentiate into other cell types. The behavior of sponge archaeocytes was compared by Sogabe et al. (2019) to that of the holozoans. This group of unicellulars is recognized as the sister group of metazoans (Lang et al., 2002), which is why the characterization of cell types and their possible changes has been the subject of careful studies in three species representing as many holozoan lineages: the choanoflagellate *Salpingoeca rosetta* (Dayel et al., 2011; Fairclough et al., 2013), the filasterean *Capsaspora owczarzaki* (Sebé-Pedrós et al., 2013, 2016), and the ichthyosporean *Creolimax fragrantissima* (de Mendoza et al., 2015). Well-characterized cell states are recognized in the life cycles of these unicells, and the transitions between them depend on mechanisms of regulation of gene expression that have been described as similar to those of metazoans (Brunet and King, 2017; Sebé-Pedrós et al., 2017).

Some authors (for example, Cembrowski and Menon [2018] for neural cells) have described the variation in the expression, within certain cell types, of some genes, including those that code for transcription factors, as continuous, and this leads to difficulties in establishing a clear dividing line between cell type and cell state (Arendt et al., 2019). A demarcation criterion could be provided (Poulin et al., 2016; Cembrowski and Menon, 2018; Tasic, 2018) by the reversibility of the transitions between states within the same cell type, contrasting with the irreversibility of terminal differentiation into different cell types. This idea is received by Arendt et al. (2019), according to whom, in practice, the different cell types could be recognized based on their ability to autonomously maintain a specific pattern of gene expression, without the need for this to be supported by a constant input from other cells or from the external environment.

1.3 HOMOLOGY PROBLEMS

Some authors who have proposed models of evolution of cell types seem to presuppose a standard catalog of functions, generally performed by distinct cell types, which, however, can sometimes be performed by one multifunctional cell type. This condition would represent a more primitive condition; from this it is presumed that a specialization took place, with division of labor between two distinct cell types (e.g., Wagner, 2019). In this model, it would be possible to recognize as "sister cell types" two types that derive from an ancestral cell type by subfunctionalization. In some cases, e.g. the cnidocytes of cnidarians and the colloblasts of comb jellies, there would actually be neofunctionalization, when one of the two sister types retains the function of the original cell type, while a new function appears with the other.

However, many zoological groups are characterized by cell types that have no morphological or functional equivalent in other animals, and it is not certain that their origin can be satisfactorily explained only in terms of *Funktionswechsel* of pre-existing cell types. In any case, a comparison between cell types of different animals opens up sensible problems of homology.

There is no risk of ambiguity using the same term (wing) to indicate the appendages that allow birds and bats to fly, even if these are homologous as forelimbs of tetrapod vertebrates but not as wings. However, we do not always have clear ideas on whether or not two structures are homologous and, therefore, on what it actually means to call them by the same name. This, of course, also applies to cell types.

Discussing this topic, Arendt et al. (2016) refer to Owen's classic definition of homology according to which a homologue is "the same organ in different animals under every variety of form and function" (Owen, 1843: 379). According to these AA., it is possible to recognize homologies between cell types, under every variety of form and function, if one accepts their evolutionary definition of cell type, "as a set of cells in an organism that change in evolution together, partially independent of other cells" (p. 744).

Similarity between two cells is not necessarily proof of their homology. This is true both for similarity at the morphological and/or functional level and similarity at the level of the expressed GRNs. It is only on the latter, however, that most recent authors rely to formulate hypotheses of homology between cell types. As a consequence, cells strictly similar in function and morphology may well emerge as the product of convergent evolution. Arendt et al. (2016) offer the example of the striated muscles of vertebrates and *Drosophila* (Brunet et al., 2016). But to find examples of convergence, it is not necessary to make comparisons between cells of animal species belonging to different phyla. Indeed, examples of convergence in cellular differentiation can be found within one and the same individual. Graham (2010) offers the example of the sensory neurons of vertebrates. The neurons of the dorsal root ganglia in the trunk perform a similar function to those of the trigeminal ganglion in the head but have distinct developmental histories.

Another phenomenon that can obscure homology between cell types is concerted evolution (Musser and Wagner, 2015; Liang et al., 2018), which results from mutations affecting genetic information expressed by multiple cell types. In this case, different cell types in the same animal can have patterns of gene expression more similar to each other than cells of the same cell type in two different species, as observed in the cells of the bioluminescent organs of two species of squid (Pankey et al., 2014). It is reasonable to speak of homology, however, when two cells of the same type of different animals turn out to be more similar to each other than two cells of different types in each of the species when compared, as found in the cells of differentiated tissues of several mammalian species (Brawand et al., 2011; Merkin et al., 2012).

1.4 CELL TYPE AS A NOMADIC CONCEPT

With reference to the transcriptional cell types recognizable in the nervous system, Zeng and Sanes (2017) and Tasic (2018) state that these can be framed in a hierarchical classification, with subtypes etc., in a kind of Linnaean taxonomy. This would be a consequence of an evolutionary history along which the underlying regulatory mechanisms have progressively diversified (Arendt, 2008; Achim et al., 2018; Arendt et al., 2019).

According to Arendt et al. (2016), cell types have the potential for independent evolutionary change. More explicitly, according to Wagner (2019: 10) "Cell types are not only functionally and developmentally distinct but also evolutionarily individualized, i.e., they can be found in different species and each has its own evolutionary history." This is not intended as a peculiarity of cell types: indeed, "[a]ny character that can be homologized is assumed to have continuity in terms of its existence in a lineage of descent, as well as persistence of differences from other parts of the body (individuality)" (Wagner, 2014: 42–43). The idea that characters can "remain themselves" throughout an indefinite number of possible alternative states is probably based on an idealistic interpretation of how organisms develop and evolve (Minelli et al., 2006); instead, an increasing number of comparative morphologists and evolutionary biologists (e.g., Roth, 1984; Haszprunar, 1992; Shubin and Wake, 1996; Minelli, 1998, 2016; Pigliucci, 2001; Minelli and Fusco, 2013) accept that evolutionary change is a continuous process, based on the remolding of preexisting features and changes in the genetic networks that regulate and control their development. Homology should be therefore treated as relative or partial.

The concept of cell type as an evolutionary unit is accepted, somehow apodictically, as the foundation for the following steps. First, "For cell types, being an evolutionary unit necessarily implies that some genomic information exists that is used only by the cells of a given type and not by other cells." (Arendt et al., 2016: 745). Second, "When we adopt this concept of cell type, cell type evolution can be equated with the evolution of the CRN/CRC [= the core regulatory network at the level of genome and the corresponding core regulatory complex at the level of transcription factors]" (Kin, 2015: 657). Third, this implies "the decoupling between cell type identity and cellular phenotypes" (Kin, 2015: 657). Definitely, the concept of cell type implicit in these statements is very different from the original one.

Eventually, any definition of cell types will depend on the subjective choice of phenotypic aspects on which our interest is focused (Valentine, 2003), especially if with phenotype we include the patterns of expressed genes and also the epigenetic modifications of genome and chromatin (Ludwig and Bintu, 2019; Morris, 2019; Wagner, 2019). At the very least, focus on epigenetic mechanisms could be useful in cases where the number of transcription factors that appear to control each specific transcriptional state are few, suggesting an important role for epigenetic factors, as in the case of ctenophores (Sebé-Pedrós et al., 2018).

It is therefore evident that the notion of cell type belongs to the category of *nomadic concepts*—concepts that have been taking different meaning and domain of application along their "migration" to new contexts (Stengers, 1987; Surman et al., 2014; Minelli, 2020b). Other examples in biology are species, gene, individual, homology.

1.5 A TENTATIVE CONCLUSION: CELL TYPE AS UNIT OF REPRESENTATION

The efforts to redefine old but eventually nomadic concepts such as species, gene, homology, or cell type show that these are still considered useful, despite the difficulties in defining and applying them in practice. As Klein (2017: 256) writes, "No single attribute has served for cell type classification. Yet "we know it when we see it." We are left with a functional but flawed taxonomy: functional, because it provides a language to describe biology; yet flawed, because it lacks consistency.

For some authors, cell types are real: "if cell types are real, then they should reflect some stable differences in the reaction norm of their gene expression" (Wagner, 2014: 252). What the stability of a cell type may be is problematic, because, as an aspect of the organization of living beings, it is subject to all the vagaries of evolution: a story of continuous transformations, in which it is necessarily arbitrary to identify in a precise event the origin of a biological structure and often impossible to give a precise meaning to the conservation of its identity over time.

In the previous pages, we have seen both the lack of a shared definition of cellular type and the uncertainties or ambiguities that often make this notion difficult to use. These difficulties have been expressed many times.

Let's start with moderate positions. According to Shendure and Trapnell (2017), the concept of cellular type is slippery but nevertheless useful enough to deserve a serious effort to obtain a satisfactory operational definition. Even when limited to the human species, as is the concern of these authors, it is in any case difficult to distinguish between cell types, insofar as they are collections of cellular states reversibly connected by possible transitions. This is also acknowledged by Kin (2015).

More drastic positions have been expressed by Elowitz (2017) and Rafelski (2017): according to these authors, we are in sight of a paradigm shift, capable of incorporating both cellular properties that vary in a discrete manner and those that vary continuously and overcoming the rigidity of the hierarchical schemes (types and subtypes) still in use today.

Finally, according to Morris (2019), criticism concerns not only the traditional morphological and functional criteria but extends to the molecular ones; transcriptomes change too rapidly in relation to the conditions outside the cell, so we cannot know if a novel transcriptional signature represents a new cell type rather than a hitherto unrecognized state of a known cell type.

Sanes (2017) rightly observes that recognizing cell types is a problem of classification. Therefore, it might be useful to look at biological systematics, a branch of life sciences with a long tradition in matters of definition and practical delimitation of the units of classification—in that case, species.

A comparison between the classification of cell types and the classification of species is legitimate and illuminating, but it becomes useful only if we focus on the more constructive aspects of the debate on the biological species. A good path, in this sense, has been traced by O'Hara (1993): "Perhaps the species problem is not something that needs to be solved, but rather something that ... needs to be gotten over" (O'Hara, 1993: 232).

As in the case of biological species (O'Hara, 1993), in addressing the "cell type problem" we do not face a problem of fact but rather a problem of historical representation. Historical, in our case, because of evolution but still more because of the transformations, reversible or not, that cells undergo in their morphology, functions, and patterns of gene expression throughout an animal's ontogeny. Again similar to species, researchers in different fields of biology and biologists dealing with different kinds of organisms (different kinds of metazoans, for the purpose of this book) may have different needs and different research agendas that may require prioritizing different properties of cells. As in biological systematics, universal generalizations of taxonomic criteria may well be impossible. This does not imply that we must abandon our efforts but rather than we must live to accept some degree of taxonomic pluralism. If our current taxonomic approach fails to answer a certain purpose, "then the way is … to create … less generalized … representations to show the features desired" (O'Hara, 1993: 244). This may require adjustments in nomenclature, into which I cannot venture at this stage, but see Minelli (2020a) for the corresponding situation in biological systematics.

REFERENCES

Achim, K., Eling, N., Vergara, H. M. et al. 2018. Whole-body single-cell sequencing reveals transcriptional domains in the annelid larval body. *Mol. Biol. Evol.* 35:1047–62.

Alié, A., Hiebert, L., Scelzo, M. and Tiozzo, S. 2020. The eventful history of non-embryonic development in colonial tunicates. *J. Exp. Zool. Part B (Mol. Dev. Evol.).* doi:10.1002/jez.b.22940.

Arendt, D. 2008. The evolution of cell types in animals: Emerging principles from molecular studies. *Nat. Rev. Genet.* 9:868–82.

Arendt, D., Bertucci, P. Y., Achim, K. and Musser, J. M. 2019. Evolution of neuronal types and families. *Curr. Opin. Neurobiol.* 56:144–52.

Arendt, D., Musser, J. M., Baker, C. V. H. et al. 2016. The origin and evolution of cell types. *Nat. Rev. Genet.* 17:744–57.

Blainey, P. 2017. Dynamic cellular personalities. In Clevers, H., Rafelski, S., Elowitz, M. et al. What is your conceptual definition of "cell type" in the context of a mature organism? *Cell Syst.* 4:258.

Borriello, E., Walker, S. I. and Laubichler, M. D. 2020. Cell phenotypes as macrostates of the GRN dynamics. *J. Exp. Zool. Part B (Mol. Dev. Evol.)* 334:213–24.

Brawand, D., Soumillon, M., Necsulea, A. et al. 2011. The evolution of gene expression levels in mammalian organs. *Nature* 478:343–8.

Brunet, T. and King, N. 2017. The origin of animal multicellularity and cell differentiation. *Dev. Cell* 43:124–40.

Brunet, T., Fischer, A. H., Steinmetz, P. R., Lauri, A., Bertucci, P. and Arendt, D. 2016. The evolutionary origin of bilaterian smooth and striated myocytes. *eLife* 5:e19607.

Cao, J., Spielmann, M., Qiu, X. et al. 2019. The single-cell transcriptional landscape of mammalian organogenesis. *Nature* 566:496–502.

Cembrowski, M. S. and Menon, V. 2018. Continuous variation within cell types of the nervous system. *Trends Neurosci.* 41:337–48.

Davidson, E. H. and Erwin, D. H. 2006. Gene regulatory networks and the evolution of animal body plans. *Science* 311:796–800.

Dayel, M. J., Alegado, R. A., Fairclough, S. R. et al. 2011. Cell differentiation and morphogenesis in the colony-forming choanoflagellate *Salpingoeca rosetta*. *Dev. Biol.* 357:73–82.

de Mendoza, A., Suga, H., Permanyer, J., Irimia, M. and Ruiz-Trillo, I. 2015. Complex transcriptional regulation and independent evolution of fungal-like traits in a relative of animals. *eLife* 4:e08904.

Eberwine, J. 2017. Are all cells unicorns? In Clevers, H., Rafelski, S., Elowitz, M. et al. What is your conceptual definition of "cell type" in the context of a mature organism? *Cell Syst.* 4:258.

Elowitz, M. 2017. Cellular demographies. In Clevers, H., Rafelski, S., Elowitz, M. et al. What is your conceptual definition of "cell type" in the context of a mature organism? *Cell Syst.* 4:255.

Fairclough, S. R., Chen, Z., Kramer, E. et al. 2013. Premetazoan genome evolution and the regulation of cell differentiation in the choanoflagellate *Salpingoeca rosetta*. *Genome Biol.* 14:1–15.

Graf, T. and Enver, T. 2009. Forcing cells to change lineages. *Nature* 462:587–94.

Graham, A. 2010. Developmental homoplasy: Convergence in cellular differentiation. *J. Anat.* 216:651–5.

Haszprunar, G. 1992. The types of homology and their significance for evolutionary biology and phylogenetics. *J. Evol. Biol.* 5:13–24.

Hobert, O. 2011. Regulation of terminal differentiation programs in the nervous system. *Annu. Rev. Cell Dev. Biol.* 27:681–96.

Kin, K. 2015. Inferring cell type innovations by phylogenetic methods—concepts, methods, and limitations. *J. Exp. Zool. Part B (Mol. Dev. Evol.)* 324B:653–61.

Klein, A. 2017. Farewell, "cell type." In Clevers, H., Rafelski, S., Elowitz, M. et al. What is your conceptual definition of "cell type" in the context of a mature organism? *Cell Syst.* 4:256.

Lang, B. F., O'Kelly, C., Nerad, T., Gray, M. W. and Burger, G. 2002. The closest unicellular relatives of animals. *Curr. Biol.* 12:1773–8.

Liang, C., Musser, J. M., Cloutier, A., Prum, R. O. and Wagner, G. P. 2018. Pervasive correlated evolution in gene expression shapes cell and tissue type transcriptomes. *Genome Biol. Evol.* 10:538–52.

Ludwig, C. H. and Bintu, L. 2019. Mapping chromatin modifications at the single cell level. *Development* 146:dev170217.

Lundberg, E. and Uhlen, M. 2017. Mapping as a key first step. In Clevers, H., Rafelski, S., Elowitz, M. et al. What is your conceptual definition of "cell type" in the context of a mature organism? *Cell Syst.* 4:257.

Manni, L. and Burighel, P. 2006. Common and divergent pathways in alternative developmental processes of ascidians. *BioEssays* 28:902–12.

Matthews, K. R. 2005. The developmental cell biology of *Trypanosoma brucei*. *J. Cell Sci.* 118:283–90.

Merkin, J., Russell, C., Chen, P. and Burge, C. B. 2012. Evolutionary dynamics of gene and isoform regulation in mammalian tissues. *Science* 338:1593–9.

Minelli, A. 1998. Molecules, developmental modules and phenotypes: A combinatorial approach to homology. *Mol. Phylogenet. Evol.* 9:340–7.

Minelli, A. 2003. *The Development of Animal Form: Ontogeny, Morphology, and Evolution*. Cambridge: Cambridge University Press.

Minelli, A. 2016. Tracing homologies in an ever-changing world. *Riv. Estet.* 62:40–55.

Minelli, A. 2020a. Taxonomy needs pluralism, but a controlled and manageable one. *Megataxa* 1:9–18.

Minelli, A. 2020b. Disciplinary fields in the life sciences: Evolving divides and anchor concepts. *Philosophies* 5, 34.

Minelli, A. and Fusco, G. 2013. Homology. In Kampourakis, K. (Ed.), *The Philosophy of Biology: A Companion for Educators*. Dordrecht: Springer: 289–322.

Minelli, A., Negrisolo, E. and Fusco, G. 2006. Reconstructing animal phylogeny in the light of evolutionary developmental biology. In Hodkinson, T. R., Parnell, J. A. N. and Waldren, S. (Eds.), *Reconstructing the Tree of Life: Taxonomy and Systematics of Species Rich Taxa* (Systematics Association Special Series, Vol. 72). Boca Raton, FL: Taylor & Francis/CRC Press: 177–90.

Morris, S. A. 2019. The evolving concept of cell identity in the single cell era. *Development* 146:dev169748.

Musser, J. M. and Wagner, G. P. 2015. Character trees from transcriptome data: Origin and individuation of morphological characters and the so-called "species signal." *J. Exp. Zool. Part B (Mol. Dev. Evol.)* 324B:588–604.

O'Hara, R. J. 1993. Systematic generalization, historical fate, and the species problem. *Syst. Biol.* 42:231–46.

Owen, R. 1843. *Lectures on the Comparative Anatomy and Physiology of the Invertebrate Animals, Delivered at the Royal College of Surgeons.* London: Longman Brown Green and Longmans.

Pankey, M. S., Minin, V. N., Imholte, G. C., Suchard, M. A. and Oakley, T. H. 2014. Predictable transcriptome evolution in the convergent and complex bioluminescent organs of squid. *Proc. Natl. Acad. Sci. USA* 111:E4736–42.

Pigliucci, M. 2001. Characters and environments. In Wagner, G. P. (Ed.), *The Character Concept in Evolutionary Biology*. San Diego, CA: Academic Press: 363–88.

Poulin, J. F., Tasic, B., Hjerling-Leffler, J., Trimarchi, J. M. and Awatramani, R. 2016. Disentangling neural cell diversity using single-cell transcriptomics. *Nat. Neurosci.* 19:1131–41.

Rafelski, S. 2017. Defining cell type space. In Clevers, H., Rafelski, S., Elowitz, M. et al. What is your conceptual definition of "cell type" in the context of a mature organism? *Cell Syst.* 4:255.

Richter, D. J. and King, N. 2013. The genomic and cellular foundations of animal origins. *Annu. Rev. Genet.* 47:509–37.

Roth, V. L. 1984. On homology. *Biol. J. Linn. Soc.* 22:13–29.

Sanes, J. R. 2017. Moving forward despite quarrels. In Clevers, H., Rafelski, S., Elowitz, M. et al. What is your conceptual definition of "cell type" in the context of a mature organism? *Cell Syst.* 4:257.

Schierwater, B. and DeSalle, R. 2018. Placozoa. *Curr. Biol.* 28:R97–8.

Sebé-Pedrós, A., Ballaré, C., Parra-Acero, H. et al. 2016. The dynamic regulatory genome of *Capsaspora* and the origin of animal multicellularity. *Cell* 165:1224–37.

Sebé-Pedrós, A., Chomsky, E., Pang, K. et al. 2018. Early metazoan cell type diversity and the evolution of multicellular gene regulation. *Nat. Ecol. Evol.* 2:1176–88.

Sebé-Pedrós, A., Degnan, B. M. and Ruiz-Trillo, I. 2017. The origin of Metazoa: A unicellular perspective. *Nat. Rev. Genet.* 18:498–512.

Sebé-Pedrós, A., Irimia, M., del Campo, J. et al. 2013. Regulated aggregative multicellularity in a close unicellular relative of metazoa. *eLife* 2:e01287.

Shendure, J. and Trapnell, C. 2017. *C. elegans* is a test-bed for ideas. In Clevers, H., Rafelski, S., Elowitz, M., et al. What is your conceptual definition of "cell type" in the context of a mature organism? *Cell Syst.* 4:256.

Shubin, N. and Wake, D. 1996. Phylogeny, variation and morphological integration. *Am. Zool.* 36:51–60.

Smith, C. L., Varoqueaux, F., Kittelmann, M. et al. 2014. Novel cell types, neurosecretory cells, and body plan of the early-diverging metazoan *Trichoplax adhaerens*. *Curr. Biol.* 24:1565–72.

Sogabe, S., Hatleberg, W. L., Kocot, K. M. et al. 2019. Pluripotency and the origin of animal multicellularity. *Nature* 570:519–22.

Stengers, I. (Ed.) 1987. *D'une science à l'autre: des concepts nomades.* Paris: Seuil.

Surman, J., Stráner, K. and Haslinger, P. 2014. Nomadic concepts—biological concepts and their careers beyond biology. *Contr. Hist. Concepts* 9(2):1–17.

Tasic, B. 2018. Single cell transcriptomics in neuroscience: Cell classification and beyond. *Curr. Opin. Neurobiol.* 50:242–9.

Valentine, J. W. 2003. Cell types, numbers, and body plan complexity. In Hall, B. K. and Olson, W. M. (Eds.), *Keywords and Concepts in Evolutionary Developmental Biology*. Cambridge, MA: Harvard University Press.

Virchow, R. 1858. *Die Cellularpathologie in ihrer Begründung auf physiologische und pathologische Gewebenlehre.* Berlin: Hirschwald.

Wagner, G. P. 2007. The developmental genetics of homology. *Nat. Rev. Genet.* 8:473–9.

Wagner, G. P. 2014. *Homology, Genes and Evolutionary Innovation.* Princeton, NJ: Princeton University Press.

Wagner, G. P. 2019. Devo-evo of cell types. In Nuño de la Rosa, L. and Müller, G. B. (Eds.), *Evolutionary Developmental Biology.* Cham: Springer Nature Switzerland.

Xia, B. and Yanai, I. 2019. A periodic table of cell types. *Development* 146:dev169854.

Zeng, H. and Sanes, J. R. 2017. Neuronal cell-type classification: Challenges, opportunities and the path forward. *Nat. Rev. Neurosci.* 18:530–46.

2 The Protistan Origins of Animal Cell Differentiation

Sebastián R. Najle and Iñaki Ruiz-Trillo

CONTENTS

2.1 Introduction 13
2.2 Cell Differentiation Is an Attribute of Animal Multicellularity..................... 13
2.3 Transdifferentiation Capacities in Animals and Other Eukaryotes 14
2.4 Unicellular Holozoans Have Complex Life Cycles That Involve Tightly Regulated Temporal Cell States 16
2.5 The Unicellular Relatives of Animals May Hold the Key to Unraveling the Origin of Animal Cell Types 21
References 22

2.1 INTRODUCTION

The first important transition during animal evolution was the transition to multicellularity from a unicellular common ancestor (Nielsen, 2008). One critical step during this transition was the origin of specialized cell types, which are the functional units of animal multicellularity. To understand how this transition occurred, we need to know first what the most recent unicellular ancestor of metazoans was like. In recent years, progress has been made in reconstructing the genomic nature of this unicellular ancestor of animals, mostly by comparing the genomes of metazoans with the genomes of their closest extant unicellular relatives. We now know that the unicellular ancestor had a complex repertoire of genes for cell adhesion, cell signalling, and transcriptional regulation. So, besides gene innovation, co-option of genes into new functions and an increased genomic regulation must have played a vital role in the evolution of animals and hence in the emergence of animal cell types. In this chapter, we review the current state of knowledge on the biology of the closest extant unicellular relatives to metazoans and propose them as suitable study models to understand the origin of animal cell types.

2.2 CELL DIFFERENTIATION IS AN ATTRIBUTE OF ANIMAL MULTICELLULARITY

The leading hypothesis to explain the origin of complex multicellularity in animals is the so-called "choanoblastea theory," which suggests that animals evolved from a unicellular ancestor that did not possess differentiated cell types through a stage of simple, undifferentiated multicellularity. This unicellular ancestor was hypothesized as a clonally dividing organism, similar to choanoflagellate colonies, that subsequently evolved specialized cell types (King, 2004; Nielsen, 2008). However, this theory has been challenged by the recent findings on animals' closest unicellular relatives. The competing hypothesis suggests that complex multicellular animals originated from the spatiotemporal integration of pre-existing temporal cell states and their associated molecular mechanisms (Mikhailov et al., 2009). Even though these two contrasting hypotheses are not seemingly mutually exclusive, there is an important difference between them, which is whether spatial cell differentiation appeared before or after the origin of multicellularity. In the first case, spatial cell differentiation and animal-like mechanisms of cell differentiation are assumed to have evolved after (or at the same time as) the acquisition of multicellularity. On the contrary, the second hypothesis implies that cell differentiation preceded multicellularity.

Whether spatial cell differentiation did or did not precede animal multicellularity is a crucial aspect if one wants to understand animal origins and the origins of animal cell types. Indeed, an essential attribute of the complex multicellularity of animals is the cooperative association of a broad spectrum of cell types. If we define cell types as cells with unique physiological and/or structural characteristics (Arendt, 2008), even the earliest branching metazoan phyla (i.e., poriferans, ctenophores, placozoans, and cnidarians) are built by morphologically and physiologically different cell types, specialized in performing specific functions.

Sponges, for instance, which are considered by some authors to be the sister clade to all other animals (Simion et al., 2017; King and Rokas, 2017), are mostly formed by a handful of well-defined cell types, including pinacocytes, choanocytes, and, depending on the species, a varying selection of mesohyl cell types, like archaeocytes and sclerocytes (recently reviewed in Adamska, 2018). Likewise, placozoans, considered the simplest known multicellular animals, are formed by six basic, morphologically defined cell types, including ciliated epithelial cells, gland cells, fiber cells, lipophil cells, crystal cells, and an array of peptidergic cells (Smith et al., 2014). Thus, highly specialized and spatially regulated cell types are an important attribute of animals.

2.3 TRANSDIFFERENTIATION CAPACITIES IN ANIMALS AND OTHER EUKARYOTES

Choanocytes (collar cells) are the most abundant cell type in sponges, and they form the choanoderm, in which they propel water and capture food. Choanocytes have a collar complex formed by an actin-based microvilli collar surrounding a motile flagellum that beats, propelling water. The collar complex is not only restricted to sponges but widely distributed in other animal lineages. Some authors consider the collar complex from metazoans to be homologous to the collar from choanoflagellates, this putative homology being the basis for the choanoblastea theory mentioned above. Therefore, this structure has been many times suggested as a synapomorphy of the choanozoan lineage, the clade formed by choanoflagellates and metazoans

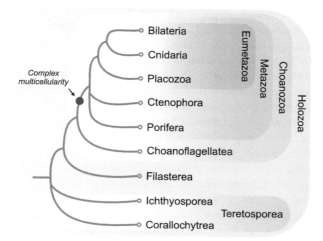

FIGURE 2.1 A phylogenetic tree that show the relationships between unicellular holozoans and animals. A dark circle depicts the evolutionary origin of animal complex multicellularity.

(Figure 2.1) (recently reviewed by Brunet and King, 2017). The potential actual homology of choanoflagellates and sponge choanocytes has, however, been put into question, first, by a morphological analysis that described differences in the morphology of the collar and in the interactions between the collar and flagella (Mah, Christensen-Dalsgaard, and Leys, 2014). Moreover, a recent study that compared the transcriptional profiles of different cell types from the sponge *Amphimedon queenslandica* among themselves and with the different life stages of the colonial choanoflagellate *Salpingoeca rosetta* showed that choanocytes are indeed the less similar sponge cell type to choanoflagellates (Sogabe et al., 2019). Nonetheless, these results should be taken with care, given that comparison of global gene expression profiles might not doubtlessly uncover cell type homologies (Liang et al., 2018).

Regardless of whether the collar of sponge choanocytes and choanoflagellates are homologous, there is one more interesting feature about sponges worth mentioning here. This is the remarkable capacity for regeneration of some sponge species. This extraordinary capacity was attributed to the highly dynamic motility behavior observed in their component cells and to the unusual plasticity of cell differentiation, which allows an exceptional efficiency for regeneration (Wilson, 1907; Simpson, 1984; James-Clark, 1867; Kent, 1880). Several lines of evidence have suggested that cell differentiation in sponges is labile, with morphologically defined cell types transdifferentiating between one another, especially during regeneration (Nakanishi, Sogabe, and Degnan, 2014; Borisenko et al., 2015; Funayama, 2018; Ereskovsky et al., 2019; Sogabe et al., 2019). The lability of cell differentiation observed in sponges prompted the idea that sponges are formed by cells at different temporal "states," rather than by terminally differentiated cell types (Adamska, 2018). This is, however, not an exclusive character of sponges. Transdifferentiation between cell types of different developmental origin also occurs during tissue regeneration in other invertebrates, like echinoderms (Mashanov,

Dolmatov, and Heinzeller, 2005; Kondo and Akasaka, 2010; Kalacheva et al., 2017) or cnidarians (Gahan et al., 2016). There are also examples of naturally occurring transdifferentiation during development in the round worm *Caenorhabditis elegans* (reviewed in Vibert, Daulny and Jarriault, 2018).

Transitions between different cell states, however, is not exclusive to multicellular organisms. Many unicellular eukaryotes (hereafter "protists") temporally metamorphose into different, specialized, cell types during their life cycles. What is more, they can also show cellular transdifferentiation during regeneration of multicellular structures, as recently shown in the social amoeba *Dictyostelium discoideum* (Mohri, Tanaka, and Nagano, 2020).

As explained below, most of the known extant unicellular relatives of metazoans transition through different temporal cell states along their life cycles. Remarkably, it has been observed that, in some species, the transition between two temporal cell states is bi-directional, meaning that these unicells can transdifferentiate between two different cell states in a reversible manner, depending on external cues. This suggests that the most recent unicellular ancestor of all holozoans could have had relaxed cell differentiation. The detailed study of the biology of the unicellular holozoans, as well as the understanding of the molecular and cellular processes that regulate their temporal cell states, provides an excellent opportunity to reconstruct the nature of the unicellular ancestor of animals.

2.4 UNICELLULAR HOLOZOANS HAVE COMPLEX LIFE CYCLES THAT INVOLVE TIGHTLY REGULATED TEMPORAL CELL STATES

Recent phylogenomic analyses have shown that there are three major unicellular lineages closely related to animals: the choanoflagellates, the filastereans, and the teretosporeans (ichthyosporeans and corallochytreans) (see Figure 2.1). Interestingly, those three lineages exhibit very different developmental modes and life cycles (depicted in Figure 2.2).

Choanoflagellates are free-living, bacterivorous, flagellate protists, some of which can develop into colonies (King, 2005). There are around 360 described choanoflagellate species (Richter and Nitsche, 2017), most of them marine but also from freshwater and terrestrial environments. Their cell is of 3–10 μm in diameter, and they all bear a single posterior flagellum enclosed by a collar of thin microvilli. They can have different phenotypes, including silica-based outer coverings that can be very intricate.

Filasterea is a recently described clade based on molecular data (Salchian-Tabrizi et al., 2008). They are filopodial amoebas or amoeboflagellates (depending on the species) whose size ranges from 3–5 μm to 7–14 μm in diameter. Some filastereans have the capacity to form multicellular structures by cell aggregation (Sebé-Pedrós et al., 2013). There are currently only four known species of filastereans: *Capsaspora owczarzaki*, which was described as a symbiont of the freshwater snail *Biomphalaria glabrata*; the free-living marine *Ministeria vibrans* which eats bacteria; and the recently described freshwater eukaryovores *Pigoraptor chileana* and *Pigoraptor vietnamica* (Hehenberger et al., 2017; Tikhonenkov et al., 2020).

The Protistan Origins of Animal Cell Differentiation

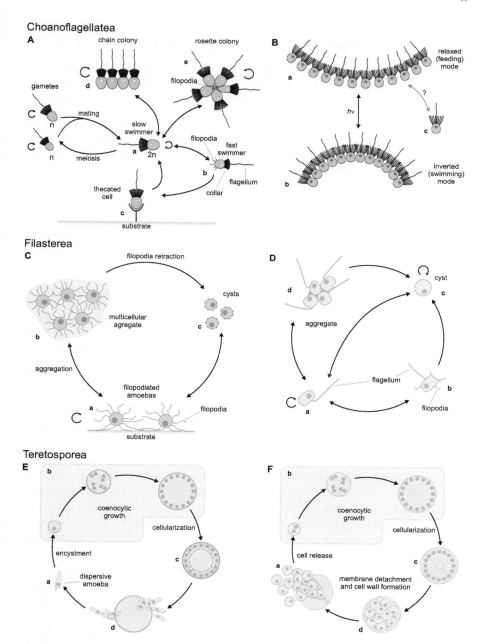

FIGURE 2.2 Schematic representation of unicellular holozoans showing different types of simple multicellularity.

FIGURE 2.2 (Continued) A. Life cycle of the colonial choanoflagellate *Salpingoeca rosetta*, including three single-celled stages: slow swimmers (**a**), fast swimmers (**b**), and thecated cells (**c**) as well as two multicellular stages: chain colonies (**d**) and rosettes (**e**). Also depicted is the sexual cycle which includes the production of haploid gametes (n), which is induced by starvation, and the reconstitution of the diploid cells (2n) by mating. **B.** Multicellular colonies of the choanoflagellate *Choanoeca flexa* transition between two behavioral modes in response to light, a relaxed mode (**a**) in which the flagella are oriented inwards and the colonies swim slowly allowing the cells to feed efficiently, and an inverted mode (**b**) with the flagella oriented outwards, in which the cells stop feeding and the colony swims faster. Whether these colonies give rise to (or are formed from) single-celled stages (**c**) has not being described. **C.** The life cycle of the filasterean *Capsaspora owczarzaki* includes the transition between three cell stages: proliferative filopodiated amoebas (**a**) can either retract their filopodia and become cystic (**c**) or actively aggregate forming multicellular clusters (**b**) with the production of an extracellular matrix of unknown composition. **D.** Reconstruction of a putative life cycle of the filasterean *Pigoraptor vietnamica*. This recently described species can show different cell morphologies including flagellated (**a**) and amoeboflagellated (**b**) cells; these can transition into rounded cysts which are proliferative (**c**). Flagellated cells were observed forming multicellular clusters (**d**), especially when preying on other unicellular eukaryotes. **E-F.** Coenocytic life cycles of the ichthyosporeans *Creolimax fragrantissima* (**E**) and *Sphaeroforma arctica* (**F**). Single-nucleated cells (**a**) undergo multiple rounds of nuclear division without cytokinesis (coenocytic growth, **b**). The coenocytes then cellularize (**c**). This process involves the formation of new cell membrane surrounding each individual nucleus. In the case of *C. fragrantissima* (**E**), the daughter cells are amoebas that are released from the mother cell (**d**) and crawl until they find a place to settle, become rounded, and encyst by forming a cell wall, restarting the cycle. On the other hand, the amoeba stage is absent in *S. arctica* (**F**) because cell wall formation in the daughter cells occurs before they are released (**d**). Part **A** is based on Brunet and King, 2017; Sebé-Pedrós, Degnan, and Ruiz-Trillo, 2017; part **B** is based on Brunet et al., 2019; part **C** is based on Sebé-Pedrós et al., 2013; part **D** is based on Hehenberger et al., 2017; Tikhonenkov et al., 2020; part **E** is based on Suga and Ruiz-Trillo, 2013; Sebé-Pedrós, Degnan, and Ruiz-Trillo, 2017; part **F** is based on Dudin et al., 2019.

Finally, the teretosporeans comprise a wide diversity of free-living, parasitic, or commensal protists, some of which have very complex life cycles comprising different cell morphologies including a multinucleated spherical or ovoid-shaped coenocyte, usually with a large central vacuole and often surrounded by a thick cell wall. This group contains two clades: the ichthyosporeans and the corallochytreans. There are more than 50 species of ichthyosporeans described. Most of them have been found to be associated with aquatic and terrestrial animals, although some free-living species were also described. Some ichthyosporeans have dispersive motile flagellated zoospores. The Corallochytrea (or Pluriformea) includes two described species, *Corallochytrium limacisporum* (Raghu-kumar, 1987), which has so far been isolated from coral reefs of India and Hawaii (Torruella et al., 2015), and the eukaryovore *Syssomonas multiformis* (Hehenberger et al. 2017; Tikhonenkov et al., 2020). The life cycle of *C. limacisporum* is relatively complex. It starts with a uninucleated cell that undergoes binary cell divisions without cytokinesis forming duplets or tetrads, sometimes also going into a multinucleated coenocytic stage. Eventually, it will release amoeboid cells that will disperse (Raghu-kumar, 1987).

We here argue that unicellular holozoans do exhibit different temporal cell types as defined under morphological and molecular criteria. Moreover, the diverse modes of simple multicellularity shown by different unicellular holozoan species provide a variety of models to tackle fundamental questions on the origin of animal multicellularity and cell differentiation.

Some choanoflagellates are known to form colonies by clonal division, a feature that, at least in the species *Salpingoeca rosetta*, is triggered by specific lipids secreted by *Algoriphagus machipongonensis*, an environmental bacterium that was co-isolated with the choanoflagellate (Alegado et al., 2012; Woznica et al., 2016). In *S. rosetta*, up to five different cell stages have been described: a sessile thecate form, slow and fast flagellate swimmers, and rosette and chain-type colonies (Figure 2.2A) (Dayel et al., 2011). Interestingly, RNA-seq data showed that each cell stage has a particular transcriptomic profile, with some genes being upregulated in specific stages. For example, genes encoding septins, which are proteins involved in cytokinesis and in maintaining epithelial cell polarity, as well as some of their potential regulators, are upregulated in the colonial stages (Fairclough et al., 2013). Moreover, the recent establishment of a transfection protocol in *S. rossetta* allowed fluorescent tagging of *S. rosetta* septins, demonstrating that these proteins localize in the basal pole of *S. rosetta* cells in a similar pattern as in animal epithelia (Booth, Szmidt-Middleton, and King, 2018). Of particular relevance for the origin of animal cell types is the recent ultrastructural analysis of *S. rosetta* colonies. The 3D reconstruction of entire rosettes, from transmission electron micrographs, demonstrates the coexistence of different cell morphologies in individual colonies, which is suggestive of spatial cell differentiation (Laundon et al., 2019; Naumann and Burkhardt, 2019).

Another interesting example of a colonial choanoflagellate is the recently described *Choanoeca flexa* (Brunet et al., 2019). This species was found in splash pools in the Caribbean and forms cup-shaped multicellular colonies which can invert their curvature in response to light (Figure 2.2B). Light perception relies on a rhodopsin-cyclic guanosine monophosphate pathway, and the signal produces a collective contractile behavior in *C. flexa* colonies that leads to an inversion of the curvature. This process is mediated by an apical acto-myosin machinery and allows the cells to alter between "feeding" and "swimming" modes. This finding has implications for the evolution of animal contractility (Brunet et al., 2019).

Among the few described species of filastereans, *Capsaspora owczarzaki* is becoming a model system, and thus a significant amount of data has been generated for this species. Its life cycle includes three different stages (Figure 2.2C). First, there is a proliferative, trophic stage of amoeboid cells with actin-based filopodia. This trophic stage can differentiate either into a cystic stage, by retracting the filopodia, or into a multicellular aggregative stage formed by different amoeboid cells actively aggregating (which includes the formation of an extracellular matrix in between the cells) (Sebé-Pedrós et al., 2013). RNA-Seq analyses of the different life stages showed that they are defined by specific transcriptional profiles, with hundreds of genes differentially expressed. For example, the aggregative stage shows upregulation of genes involved in the integrin adhesome and extracellular matrix, while there is upregulation of genes involved in filopodia formation, actin cytoskeleton, or tyrosine kinases in the amoeboid proliferative stage. The study also showed differential alternative splicing and differential expression of long non-coding RNAs linked to different life stages (Sebé-Pedrós et al., 2013). Further analyses of *C. owczarzaki* showed the presence of different animal-like mechanisms of cell differentiation in this taxon. For example, transitions into the different life cycle stages are dictated by extensive remodelling of the proteome and dynamic phosphosignalling events (Sebé-Pedrós, Peña, et al., 2016). This stage-specific phosphoregulation affects dozens of cytoplasmic and receptor tyrosine kinases as well as several transcription factors whose homologs in animals are involved in development (such as NF-kappaB, Runx, CREB, and p53). Moreover, multiple functional genomics assays show that cell state transitions in *C. owczarzaki* are also dictated by changing chromatin states, differential deployment of lincRNAs, and dynamic cis-regulation (Sebé-Pedrós, Ballaré, et al., 2016).

The recent description of two new species of filastereans, *P. chileana* and *P. vietnamica*, which have similar morphologies and conspicuous cell states, further supports the idea that complex life histories and the formation of simple multicellular stages is widespread among the unicellular holozoans. *Pigoraptor spp.* are flagellates and/or amoeboflagellates, which can transition into spherical, proliferative "cysts" and can also form multicellular aggregates (Figure 2.2D) (Tikhonenkov et al., 2020). Even though their cell and molecular biologies have not yet been extensively studied, as compared with other unicellular holozoans, their genomes encode a rich repertoire of genes for cell adhesion and extracellular matrix components (Hehenberger et al., 2017).

Finally, teretosporeans can show different life cycles, depending on the species. Some species have flagellated cells; others are amoeboid, but most of the described species develop through a multinucleated coenocyte that will eventually give rise to several uninucleated daughter cells. The best-studied species are the ichthyosporeans *Creolimax fragrantissima* and *Sphaeroforma arctica*. The life cycle of *C. fragrantissima* starts with a small cell with a single nucleus and a cell wall. This cell grows in volume and develops, by synchronic nuclear divisions, into a coenocyte with dozens of nuclei. The coenocyte will then go through a cellularization process in which individual nuclei are enclosed in a newly formed cell membrane. This process will end with the release, through breaks in the external cell wall, of uninucleated amoebas. These amoebas will disperse, encyst (with the formation of an external cell wall), and start the life cycle again (Figure 2.2E). Transcriptomic analysis of the amoeba and the coenocyte life stages of *C. fragrantissima* show specific profiles for each cell stage, with some genes and functions differentially expressed. For example, most of the genes encoding proteins involved in the integrin adhesome (such as integrins, paxillin, vinculin, or parvin) are upregulated in the amoeba stage, together with genes involved in protein kinase activity or actin cytoskeleton. In contrast, genes involved in DNA replication or RNA metabolism are upregulated in the multinucleate, coenocyte stage. Moreover, that study showed differential alternative splicing among cell stages, as well as differential expression of long non-coding RNAs (de Mendoza et al., 2015). A more recent study showed the involvement of the acto-myosin cytoskeleton driving inward membrane invagination during the cellularization of the *S. arctica* coenocyte. The life cycle of *S. arctica* is very similar to the life cycle of *C. fragrantissima* (compare Figures 2.2E and F). Cellularization in *S. arctica* leads to the formation of a polarized layer of cells that localizes at the periphery of the mature mother cell and somehow resembles the epithelium of cellularized insect embryos. One difference between the *S. arctica* life cycle and that of *C. fragrantissima* is the absence of a dispersive amoeba stage. This is due to the fact that cell wall formation in *S. arctica* occurs before the daughter cells are released (Figure 2.2E). Remarkably, transcriptomic analysis showed that this stage is characterized by the upregulation of genes involved in cell adhesion (Dudin et al., 2019).

2.5 THE UNICELLULAR RELATIVES OF ANIMALS MAY HOLD THE KEY TO UNRAVELING THE ORIGIN OF ANIMAL CELL TYPES

As outlined above, an important feature of unicellular holozoans is their complex life cycles, which involve diverse temporal cell states of varying morphology and physiology, governed by tight genetic regulation. For example, in *C. owczarzaki*, transitions from one cell stage to another are regulated at the level of gene expression, alternative splicing, long non-coding RNAs, posttranslational modifications of histone tails, and also differential phosphorylation of transcription factors. Thus, *C. owczarzaki* uses animal-like mechanisms of cell differentiation to transition from one cell stage to another cell stage.

Taken together, the data on unicellular holozoans provide compelling evidence to suggest that the raw material for the evolution of cell type-specific differentiation was already present in the unicellular ancestor of Metazoa. However, how and when those gene inventories were networked and regulated to give rise to the cooperative spatiotemporal association of different cell types of the complex animal multicellularity remain fundamental questions in biology.

To answer these questions, it is imperative to investigate signatures of cell differentiation in the unicellular relatives of animals. First steps on this direction were taken by the previously mentioned studies which showed the presence, in the rosette colonies of the choanoflagellate *S. rosetta*, of different cell morphologies that display clear ultrastructural modifications (Laundon et al., 2019; Naumann and Burkhardt, 2019). Based on this observation, the authors suggested that spatial cell type differentiation might have been present in the most recent ancestor of choanoflagellates and animals. But one important limitation of this assumption is that the differentiated cell types are defined solely on morphological information. At the moment, there is no molecular data proving the spatial coexistence of cell types in the simple multicellular stages of choanoflagellates, filastereans, or ichthyosporeans. Traditionally, cell types have been characterized using both microscopy and molecular fingerprinting tools (such as in situ hybridization) that characterize the expression pattern of one or a few selected gene markers. However, given that these methods rely on the previous characterization of molecular markers of cell differentiation, they were usually limited to a handful of well-studied model species.

The recent development of single-cell sequencing methods, like single-cell RNA-seq, are revolutionizing our understanding of cell differentiation. These methods can overcome many technical limitations and have opened the way to systematically characterize the diversity of cell types in different tissues and organisms (Jaitin et al., 2014; Tasic et al., 2016; Cao et al., 2017; Sebé-Pedrós, Saudemont, et al., 2018) and also to model cell differentiation dynamics (reviewed in Griffiths et al., 2018). Importantly, single-cell genomics requires minimal conjectural knowledge on the system and, therefore, is independent of *a priori* selection of candidate marker genes. Moreover, unlike conventional high-throughput sequencing methods that measure an average reading across pooled cell populations, single-cell sequencing methods do not require prior enrichment of particular cell populations. Therefore, it is applicable to non-standard model systems, as recently shown (Sebé-Pedrós, Chomsky, et al., 2018). Thus, we believe that the next logical step towards assessing cell differentiation in the close unicellular relatives of animals is the application of single-cell RNA-seq to characterize molecularly the cellular heterogeneity within their multicellular life stages. This will allow the reconstruction of the evolutionary steps that led to the origin of animal multicellularity and cell type differentiation, as well as the ability to trace the origin of specific cellular phenotypes and transcriptional programs. Moreover, the recent establishment of transgenesis tools in *S. rosetta* (Booth, Szmidt-Middleton, and King, 2018; Wetzel et al., 2018) and *C. owczarzaki* (Parra-Acero et al., 2018) offers the unprecedented opportunity to validate the predicted cell type-specific expression signatures.

In summary, the origin of animal cell differentiation has to be sought among the closest unicellular relatives of animals. Our better understanding of their biology, together with the advent of single-cell technologies and the development of genetic tools in some of those taxa, will surely provide unique insights into this crucial question.

REFERENCES

Adamska, Maja. 2018. Differentiation and transdifferentiation of sponge cells. *Results and Problems in Cell Differentiation* 65: 229–53. doi:10.1007/978-3-319-92486-1_12.

Alegado, Rosanna A., Laura W. Brown, Shugeng Cao, Renee K. Dermenjian, Richard Zuzow, Stephen R. Fairclough, Jon Clardy, and Nicole King. 2012. A bacterial sulfonolipid triggers multicellular development in the closest living relatives of animals. *ELife* 2012 (1): 2011–13. doi:10.7554/eLife.00013.

Arendt, Detlev. 2008. The evolution of cell types in animals: Emerging principles from molecular studies. *Nature Reviews. Genetics* 9 (11): 868–82. doi:10.1038/nrg2416.

Booth, David S., Heather Szmidt-Middleton, and Nicole King. 2018. Transfection of choanoflagellates illuminates their cell biology and the ancestry of animal septins. Edited by Doug Kellogg. *Molecular Biology of the Cell* 29 (25): 3026–38. doi:10.1091/mbc.E18-08-0514.

Borisenko, Ilya E., Maja Adamska, Daria B. Tokina, and Alexander V. Ereskovsky. 2015. "Transdifferentiation Is a Driving Force of Regeneration in *Halisarca Dujardini* (Demospongiae, Porifera)." *PeerJ* 3: e1211. doi:10.7717/peerj.1211.

Brunet, Thibaut, and Nicole King. 2017. The origin of animal multicellularity and cell differentiation. *Developmental Cell* 43 (2): 124–40. doi:10.1016/j.devcel.2017.09.016.

Brunet, Thibaut, Ben T. Larson, Tess A. Linden, Mark J. A. Vermeij, Kent McDonald, and Nicole King. 2019. Light-regulated collective contractility in a multicellular choanoflagellate. *Science (New York, N.Y.)* 366 (6463): 326–34. doi:10.1126/science. aay2346.

Cao, Junyue, Jonathan S. Packer, Vijay Ramani, Darren A. Cusanovich, Chau Huynh, Riza Daza, Xiaojie Qiu, et al. 2017. Comprehensive single-cell transcriptional profiling of a multicellular organism. *Science.* doi:10.1126/science.aam8940.

Dayel, Mark J., Rosanna A. Alegado, Stephen R. Fairclough, Tera C. Levin, Scott A. Nichols, Kent McDonald, and Nicole King. 2011. Cell differentiation and morphogenesis in the colony-forming choanoflagellate *Salpingoeca rosetta. Developmental Biology* 357 (1): 73–82. doi:10.1016/j.ydbio.2011.06.003.

Dudin, Omaya, Andrej Ondracka, Xavier Grau-Bové, Arthur A. B. Haraldsen, Atsushi Toyoda, Hiroshi Suga, Jon Bråte, and Iñaki Ruiz-Trillo. 2019. A unicellular relative of animals generates a layer of polarized cells by actomyosin-dependent cellularization. *ELife* 8: 1–27. doi:10.7554/eLife.49801.

Ereskovsky, Alexander V., Daria B. Tokina, Danial M. Saidov, Stephen Baghdiguian, Emilie Le Goff, and Andrey I. Lavrov. 2019. Transdifferentiation and mesenchymal-to-epithelial transition during regeneration in Demospongiae (Porifera). *Journal of Experimental Zoology Part B: Molecular and Developmental Evolution* (October 2019): 37–58. doi:10.1002/jez.b.22919.

Fairclough, Stephen R., Zehua Chen, Eric Kramer, Qiandong Zeng, Sarah Young, Hugh M. Robertson, Emina Begovic, et al. 2013. Premetazoan genome evolution and the regulation of cell differentiation in the choanoflagellate *Salpingoeca rosetta. Genome Biology* 14 (2): R15. doi:10.1186/gb-2013-14-2-r15.

Funayama, Noriko. 2018. The cellular and molecular bases of the sponge stem cell systems underlying reproduction, homeostasis and regeneration. *International Journal of Developmental Biology* 62 (6–8): 513–25. doi:10.1387/ijdb.180016nf.

Gahan, James M., Brian Bradshaw, Hakima Flici, and Uri Frank. 2016. The interstitial stem cells in *Hydractinia* and their role in regeneration. *Current Opinion in Genetics and Development* 40: 65–73. doi:10.1016/j.gde.2016.06.006.

Griffiths, Jonathan A., Antonio Scialdone, and John C. Marioni. 2018. Using single-cell genomics to understand developmental processes and cell fate decisions. *Molecular Systems Biology* 14 (4): e8046. doi:10.15252/msb.20178046.

Hehenberger, Elisabeth, Denis V. Tikhonenkov, Martin Kolisko, Javier del Campo, Anton S. Esaulov, Alexander P. Mylnikov, and Patrick J. Keeling. 2017. Novel predators reshape holozoan phylogeny and reveal the presence of a two-component signaling system in the ancestor of animals. *Current Biology* 27 (13): 2043–50.e6. doi:10.1016/j.cub.2017.06.006.

Jaitin, D. A., E. Kenigsberg, H. Keren-Shaul, N. Elefant, F. Paul, I. Zaretsky, A. Mildner, et al. 2014. Massively parallel single-cell RNA-seq for marker-free decomposition of tissues into cell types. *Science* 343 (6172): 776–79. doi:10.1126/science.1247651.

James-Clark, H. 1867. "IV.-Conclusive Proofs of the Animality of the Ciliate Sponges, and of Their Affinities with the Infusoria Flagellate." *The Annals and Magazine of Natural History Including Zoology, Botany, and Geology* 19 (109): 13–18.

Kalacheva, Nadezhda V., Marina G. Eliseikina, Lidia T. Frolova, and Igor Yu Dolmatov. 2017. Regeneration of the digestive system in the crinoid *Himerometra robustipinna* occurs by transdifferentiation of neurosecretory-like cells. *PLoS One* 12 (7): 1–28. doi:10.1371/journal.pone.0182001.

Kent, William Saville. 1880. *A Manual of the Infusoria : Including a Description of All Known Flagellate, Ciliate, and Tentaculiferous Protozoa, British and Foreign, and an Account of the Organization and the Affinities of the Sponges.* London: D. Bogue.

King, Nicole. 2004. The unicellular ancestry of animal development. *Developmental Cell* 7 (3): 313–25. doi:10.1016/j.devcel.2004.08.010.

King, Nicole. 2005. Choanoflagellates. *Current Biology: CB* 15 (4): R113–4. doi:10.1016/j.cub.2005.02.004.

King, Nicole, and Antonis Rokas. 2017. Review embracing uncertainty in reconstructing early animal evolution. *Current Biology* 27 (19): R1081–88. doi:10.1016/j.cub.2017.08.054.

Kondo, Mariko, and Koji Akasaka. 2010. Regeneration in crinoids. *Development, Growth and Differentiation* 52 (1): 57–68. doi:10.1111/j.1440-169X.2009.01159.x.

Laundon, Davis, Ben T. Larson, Kent McDonald, Nicole King, and Pawel Burkhardt. 2019. The architecture of cell differentiation in choanoflagellates and sponge choanocytes. *PLoS Biology* 17 (4): e3000226. doi:10.1371/journal.pbio.3000226.

Liang, Cong, Jacob M. Musser, Alison Cloutier, Richard O. Prum, and Günter P. Wagner. 2018. Pervasive correlated evolution in gene expression shapes cell and tissue type transcriptomes. *Genome Biology and Evolution* 10 (2): 538–52. doi:10.1093/gbe/evy016.

Mah, Jasmine L., Karen K. Christensen-Dalsgaard, and Sally P. Leys. 2014. Choanoflagellate and choanocyte collar-flagellar systems and the assumption of homology. *Evolution and Development* 16 (1): 25–37. doi:10.1111/ede.12060.

Mashanov, Vladimir S., Igor Yu Dolmatov, and Thomas Heinzeller. 2005. Transdifferentiation in holothurian gut regeneration. *Biological Bulletin* 209 (3): 184–93. doi:10.2307/3593108.

Mendoza, Alex de, Hiroshi Suga, Jon Permanyer, Manuel Irimia, and Iñaki Ruiz-Trillo. 2015. Complex transcriptional regulation and independent evolution of fungal-like traits in a relative of animals. *ELife* 4 (October): e08904. doi:10.7554/eLife.08904.

Mikhailov, Kirill V., Anastasiya V. Konstantinova, Mikhail A. Nikitin, Peter V. Troshin, Leonid Yu Rusin, Vassily A. Lyubetsky, Yuri V. Panchin, et al. 2009. The origin of Metazoa: A transition from temporal to spatial cell differentiation. *BioEssays* 31 (7): 758–68. doi:10.1002/bies.200800214.

Mohri, Kurato, Ryodai Tanaka, and Seido Nagano. 2020. Live cell imaging of cell movement and transdifferentiation during regeneration of an amputated multicellular body of the social amoeba *Dictyostelium discoideum*. *Developmental Biology* 457 (1): 140–49. doi:10.1016/j.ydbio.2019.09.014.

Nakanishi, Nagayasu, Shunsuke Sogabe, and Bernard M. Degnan. 2014. Evolutionary origin of gastrulation: Insights from sponge development. *BMC Biology* 12: 1–9. doi:10.1186/1741-7007-12-26.

Naumann, Benjamin, and Pawel Burkhardt. 2019. Spatial cell disparity in the colonial choanoflagellate *Salpingoeca rosetta*. *Frontiers in Cell and Developmental Biology* 7 (October): 1–19. doi:10.3389/fcell.2019.00231.

Nielsen, Claus. 2008. Six major steps in animal evolution: Are we derived sponge larvae? *Evolution and Development* 10 (2): 241–57. doi:10.1111/j.1525-142X.2008.00231.x.

Parra-Acero, Helena, Núria Ros-Rocher, Alberto Perez-Posada, Aleksandra Kożyczkowska, Núria Sánchez-Pons, Azusa Nakata, Hiroshi Suga, Sebastián R. Najle, and Iñaki Ruiz-Trillo. 2018. Transfection of *Capsaspora owczarzaki*, a close unicellular relative of animals. *Development* 145 (10): dev162107. doi:10.1242/dev.162107.

Raghu-kumar, S. 1987. Occurrence of the thraustochytrid, *Corallochytrium limacisporum* gen. et sp. nov. in the coral reef lagoons of the Lakshadweep Islands in the Arabian Sea. *Botanica Marina* 30 (1985): 83–89.

Richter, Daniel J., and Frank Nitsche. 2017. Choanoflagellatea. In *Handbook of the Protists*, edited by John M. Archibald, Alastair G. B. Simpson, and Claudio H. Slamovits, 1479–96. Cham: Springer International Publishing. doi:10.1007/978-3-319-28149-0_5.

Salchian-Tabrizi, Kamran, Marianne A. Minge, Mari Espelund, Russell Orr, Torgeir Ruden, Kjetill S. Jakobsen, and Thomas Cavalier-Smith. 2008. Multigene phylogeny of choanozoa and the origin of animals. *PLoS One* 3 (5): e2098. doi:10.1371/journal.pone.0002098.

Sebé-Pedrós, Arnau, Cecilia Ballaré, Helena Parra-Acero, Cristina Chiva, Juan J. Tena, Eduard Sabidó, José Luis Gómez-Skarmeta, Luciano Di Croce, and Iñaki Ruiz-Trillo. 2016. The dynamic regulatory genome of *Capsaspora* and the origin of animal multicellularity. *Cell*. doi:10.1016/j.cell.2016.03.034.

Sebé-Pedrós, Arnau, Elad Chomsky, Kevin Pang, David Lara-Astiaso, Federico Gaiti, Zohar Mukamel, Ido Amit, Andreas Hejnol, Bernard M. Degnan, and Amos Tanay. 2018. Early metazoan cell type diversity and the evolution of multicellular gene regulation. *Nature Ecology & Evolution* 2 (7): 1176–88. doi:10.1038/s41559-018-0575-6.

Sebé-Pedrós, Arnau, Bernard M. Degnan, and Iñaki Ruiz-Trillo. 2017. The origin of Metazoa: A unicellular perspective. *Nature Reviews Genetics*. doi:10.1038/nrg.2017.21.

Sebé-Pedrós, Arnau, Manuel Irimia, Javier del Campo, Helena Parra-Acero, Carsten Russ, Chad Nusbaum, Benjamin J. Blencowe, and Iñaki Ruiz-Trillo. 2013. Regulated aggregative multicellularity in a close unicellular relative of metazoa. *Elife* 2 (2): e01287. doi:10.7554/eLife.01287.001.

Sebé-Pedrós, Arnau, Marcia Ivonne Peña, Salvador Capella-Gutiérrez, Meritxell Antó, Toni Gabaldón, Iñaki Ruiz-Trillo, and Eduard Sabidó. 2016. High-throughput proteomics reveals the unicellular roots of animal phosphosignaling and cell differentiation. *Developmental Cell* 39 (2): 186–97. doi:10.1016/j.devcel.2016.09.019.

Sebé-Pedrós, Arnau, Baptiste Saudemont, Elad Chomsky, Flora Plessier, Marie Pierre Mailhé, Justine Renno, Yann Loe-Mie, et al. 2018. Cnidarian cell type diversity and regulation revealed by whole-organism single-cell RNA-seq. *Cell* 173 (6): 1520–34. e20. doi:10.1016/j.cell.2018.05.019.

Simion, Paul, Denis Baurain, Nicole King, Gert Wo, Denis Baurain, Muriel Jager, Daniel J. Richter, and Arnaud Di Franco. 2017. A large and consistent phylogenomic dataset supports sponges as the sister group to all other animals. *Current Biology* 27 (7): 1–10. doi:10.1016/j.cub.2017.02.031.

Simpson, Tracy L. 1984. *The Cell Biology of Sponges. The Cell Biology of Sponges*. New York, NY: Springer New York. doi:10.1007/978-1-4612-5214-6.

Smith, Carolyn L., Frédérique Varoqueaux, Maike Kittelmann, Rita N. Azzam, Benjamin Cooper, Christine A. Winters, Michael Eitel, Dirk Fasshauer, and Thomas S. Reese. 2014. Novel cell types, neurosecretory cells, and body plan of the early-diverging metazoan *Trichoplax adhaerens*. *Current Biology* 24 (14): 1565–72. doi:10.1016/j.cub.2014.05.046.

Sogabe, Shunsuke, William L. Hatleberg, Kevin M. Kocot, Tahsha E. Say, Daniel Stoupin, Kathrein E. Roper, Selene L. Fernandez-Valverde, Sandie M. Degnan, and Bernard M. Degnan. 2019. Pluripotency and the origin of animal multicellularity. *Nature* 570 (7762): 519–22. doi:10.1038/s41586-019-1290-4.

Suga, Hiroshi, and Iñaki Ruiz-Trillo. 2013. Development of Ichthyosporeans sheds light on the origin of metazoan multicellularity. *Developmental Biology* 377 (1): 284–92. doi:10.1016/j.ydbio.2013.01.009.

Tasic, Bosiljka, Vilas Menon, Thuc Nghi Nguyen, Tae Kyung Kim, Tim Jarsky, Zizhen Yao, Boaz Levi, et al. 2016. Adult mouse cortical cell taxonomy revealed by single cell transcriptomics. *Nature Neuroscience* 19 (2): 335–46. doi:10.1038/nn.4216.

Tikhonenkov, Denis V., Elisabeth Hehenberger, Anton S. Esaulov, Olga I. Belyakova, Yuri A. Mazei, Alexander P. Mylnikov, and Patrick J. Keeling. 2020. Insights into the origin of metazoan multicellularity from predatory unicellular relatives of animals. *BMC Biol* 18, 39. https://doi.org/10.1186/s12915-020-0762-1.

Torruella, Guifré, Alex De Mendoza, Xavier Grau-Bové, Meritxell Antó, Mark A. Chaplin, Javier Del Campo, Laura Eme, et al. 2015. Phylogenomics reveals convergent evolution of lifestyles in close relatives of animals and fungi. *Current Biology* 25 (18): 2404–10. doi:10.1016/j.cub.2015.07.053.

Vibert, Laura, Anne Daulny, and Sophie Jarriault. 2018. Wound healing, cellular regeneration and plasticity: The elegans way. *International Journal of Developmental Biology* 62 (6–8): 491–505. doi:10.1387/ijdb.180123sj.

Wetzel, Laura A., Tera C. Levin, Ryan E. Hulett, Daniel Chan, Grant A. King, Reef Aldayafleh, David S. Booth, et al. 2018. Predicted glycosyltransferases promote development and prevent spurious cell clumping in the choanoflagellate *S. rosetta*. *Elife* 7: 1–28.

Wilson, H. V. 1907. On some phenomena of coalescence and regeneration in sponges. *Journal of Experimental Zoology* 5 (2): 245–58. doi:10.1002/jez.1400050204.

Woznica, Arielle, Alexandra M. Cantley, Christine Beemelmanns, Elizaveta Freinkman, Jon Clardy, and Nicole King. 2016. Bacterial lipids activate, synergize, and inhibit a developmental switch in choanoflagellates. *Proceedings of the National Academy of Sciences of the United States of America* 113 (28): 7894–99. doi:10.1073/pnas.1605015113.

3 Convergent Evolution of Animal-Like Organelles across the Tree of Eukaryotes

Greg S. Gavelis, Gillian H. Gile and Brian S. Leander

CONTENTS

3.1 Introduction	27
3.2 Eye-Like Organelles	29
3.3 Statocyst-Like Organelles	31
3.4 Nematocyst-Like Organelles	34
3.5 Conclusions	41
References	41

3.1 INTRODUCTION

Morphogenesis through cell-type specification has allowed for incredible anatomical complexity to arise among macroorganisms such as plants, animals, fungi, seaweeds, and kelps, despite these groups evolving multicellularity independently (Parfrey and Lahr, 2013). Much of the field of "evo-devo"—and the crux of this book—is devoted to understanding how distinct cell types evolved and are coordinated to form tissues and organs with sophisticated divisions of labor. In the history of life, this capability is a relatively recent invention, as the last common eukaryotic ancestor was unicellular (as are most eukaryotic species alive today) and necessarily performed all tasks of survival and reproduction within the confines of a single cell membrane. Of course, unicellular eukaryotes or "protists" are still capable of divisions of labor, both spatially (within specialized organelles or cell regions) and temporally (e.g. by changing cell shape over the course of the lifecycle). For instance, protist relatives of animals have multifarious lifecycles that can alternate between solitary and colonial flagellated stages (in choanoflagellates), motile amoebae, and either multicellular or syncytial spheres (e.g., ichthyosporeans), that may encyst or aggregate into multinucleate colonies (e.g., filastereans) (Sebé-Pedrós, Degnan, and Ruiz Trillo, 2017). As with metazoan tissues, each of these stages is associated with distinct gene expression patterns, and uncovering the regulation of these life-stage transitions may shed light onto the early evolution of cell type specification in animals. Thus, understanding the evolution of complexity in both animals and protists are two intertwined pursuits.

Over the course of eukaryotic evolution, cellular architecture has been expanded by the acquisition of endosymbiotic organelles (e.g., mitochondria, chloroplasts, and their genes) (Gavelis et al., 2015) as well as by the multiplication of endogenous components (e.g., via gene duplications and consequent expansions in protein complexes and pathways) (Mast et al., 2014). While animals form a single eukaryotic clade characterized by embryogenesis (which is generally patterned "top down" by master regulatory genes), protists are paraphyletic—encompassing all eukaryotic diversity excluding animals, plants, and multicellular fungi—and their developmental processes will inevitably be more diverse (Lukes, Leander, and Keeling, 2009). Beyond a handful of well-studied models, the means by which protists establish polarity, symmetry, and the size of their cells and organelles are unclear, and knowledge gaps are particularly glaring for large and/or complex protists (Marshall, 2011). Some lineages have features so ornate and overwrought that they appear very inefficient (e.g. the "kinetoplast" of *Trypanosoma* consists of thousands of interlinked rings of DNA that act as templates for iterative rounds of mitochondrial RNA editing, with no obvious benefit over systems in "normal" cells) (Lukeš, Hashimi, and Zíková, 2005; Lukes, Leander, and Keeling, 2009). Thus, not only are the principles of protist evo-devo unclear, but—as in the field of comparative eukaryotic genomics—it is difficult to determine which cell features result primarily from adaptation, versus developmental constraints, versus "constructive neutral evolution" (wherein genetic drift allows for the accretion of useless details) (Lukes, Leander, and Keeling, 2009; Wideman et al., 2019). In other words, complexity is not necessarily adaptive.

Nevertheless, many elaborate protist features and body plans are adaptive and can perform roles equivalent to what animals accomplish with multitudes of cells and tissue types. For instance, haplozoan dinoflagellates are parasites that look and behave like marine tapeworms (Leander, 2008a). Each organism includes an apical attachment apparatus consisting of a hook and a sucker, microtrich-like surface extensions to facilitate nutrient absorption, and a body composed of linearly-arranged proglottid-like segments that are shed terminally for dispersal along with the of feces the host (a maldinid polychaete or "bamboo worm"). Being only distantly related to animals, this organism represents convergent evolution, where two lineages independently acquire similar structures under similar selective pressures. What is remarkable is that *Haplozoon* is not multicellular—it is a syncytium with a single plasma membrane bounding many nuclei. One nucleus is situated in the "head," while others are arranged serially in segment-like compartments, each of which is delimited by flattened membranous sacs ("alveoli") rather than true cell boundaries (Leander et al., 2002; Rueckert and Leander, 2008; Wakeman et al., 2018). Though the molecular basis of haplozoan development is as yet unstudied, the fact that its cytoplasm is physically partitioned means that morphogenesis could be shaped by differential expression among nuclei. For instance, nuclei in the syncytial alga *Caulerpa* have been found to express certain transport regulators differentially between the leaf-like "fronds" and root-like "stolons" of this meters-long cell (Arimoto et al., 2019).

Unfortunately, it is logistically harder to study how protists achieve differentiation at finer spatial scales.

Protists are capable not only of structural convergence with multicellular body plans, but also of occupying similar niches. For example, vorticellid ciliates greatly resemble entoprocts and some rotifers, all of which are filter feeders that use radially-symmetrical ciliary arrays to capture particles and attach to substrates via an adhesive disk on a long contractile stalk (Rundell and Leander, 2010). Similarly, chonotrich ciliates resemble cycliophorans, both of which have episymbiotic lifestyles near the mouthparts of decapods (e.g., crabs and lobsters), using a ciliary feeding apparatus, stalk, and adhesive disk (Funch and Kristensen, 1995; Obst et al., 2006; Taylor et al., 1995).

The above examples represent structural similarities between the trait complexes of entire organisms, but convergent evolution most often manifests at more specific levels of organization. In the next three examples, we discuss convergent evolution between organs and organelles, using marvelously complex ciliates and dinoflagellates. These are all "non-model" organisms, for they lack sequenced genomes and tools for genetic transformation, and some have not even been cultured (e.g., warnowiid dinoflagellates), but the intricacy of these systems illustrates how much we have yet to learn about protist evo-devo. All three organisms are alveolates, a lineage that diverged from the ancestors of animals over one billion years ago (Parfrey et al., 2011). This distant common ancestry suggests that homology has played little to no role in the canalization of these outcomes (Leander, 2008a,b).

3.2 EYE-LIKE ORGANELLES

Perhaps the most famous example of convergent evolution is the independent occurrence of camera-type eyes: the lens-bearing photosensory structures found in vertebrates and many mollusks, annelids, and arthropods (Gavelis et al., 2017) (Figure 3.1). While each lineage derived its own corneas, lenses, and neural wiring for image formation, these were built on top of a shared toolkit for photoreception that they inherited from their bilaterian common ancestor (i.e., opsins hosted by either rhabdomeric or ciliary cells, which are both specified by the *Pax6* master control gene) (Gehring and Ikeo, 1999). With the possible exception of lensed eyes in cubozoan cnidarians (which are found in the absence of *Pax6* orthologs), camera-type eyes in animals reflect proximate rather than ultimate convergence, as developmental genes played an important role in the canalization of independently derived camera eyes (Leander, 2008a,b).

A case of ultimate convergence with the camera-type eyes of animals exists in the eye-like "ocelloids" of warnowiid dinoflagellates (Leander, 2008b). Ocelloids resemble the camera-type eyes of some animals (e.g., cubozoans) and—when these bizarre cells were first collected from the plankton—were even mistaken for them (Kofoid and Swezy, 1921). The ocelloid consists of a cornea-like layer, a lens, and a pigmented retinal body, albeit at a subcellular scale (Gavelis et al., 2015; Hayakawa et al., 2015) (Figure 3.1). We caution that any role of the ocelloid in phototaxis (movement relative to light) is still speculative, as warnowiids are not yet in culture and

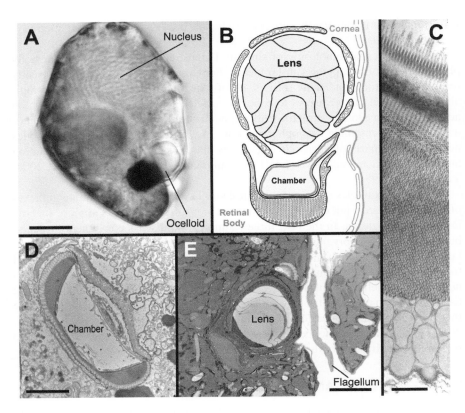

FIGURE 3.1 The eye-like ocelloid of warnowiid dinoflagellates. (**A**) A DIC light micrograph of *Nematodinium*, ocelloid visible in the lower right. (**B**) Diagram of the ocelloid, redrawn from Taylor (1980) with additional details. Red = thylakoid membranes and plastoglobules, purple = mitochondrial cristae, blue = seawater. (**C**) TEM section of the retinal body showing (from bottom to top), plastoglobules, waveform thylakoid membranes, and paracrystalline layers that may somehow serve to process incoming signals. (**D**) An oblique TEM section of the retinal body showing its concave inner surface. (**E**) FIB-SEM section of the ocelloid and longitudinal flagellum. Scale bar = 10 μm in A; 100 nm in C; 2 μm in D, 5 μm in E.

quickly become inactive when isolated from the plankton. Nevertheless, the position and refractive index of the lens suggests that the ocelloid focuses light on the inner surface of its retinal body (Francis, 1967), and while it is conceivable that the ocelloid performs non-sensory functions (e.g., serving a site for photosynthesis), most of its properties are consistent with a phototactic role (e.g. "oculomotor" swiveling motion and its homology to phototactic eyespots in other dinoflagellates, as we will discuss). Our fragmentary knowledge of the ocelloid is still remarkable from an evolutionary and developmental standpoint. Prior to each cell division, the lens, cornea, and retinal body each de-differentiate and divide by binary fission. They are then equally apportioned between daughter cells, each reassembling around an "ocellar fiber" that tethers the ocelloid to a flagellum (Dodge and Greuet, 1987).

It is simplest to describe ocelloid ultrastructure from the outside in: The cornea-like layer is composed of a network of mitochondria that send membranous projections into the surface of the lens (where new crystalline layers are accreted) and may aid in lens formation (Figure 3.1). The lens itself is comprised of concentric vesicles, where the innermost crystals are the most densely packed. At its mid-region, the lens is surrounded by either two or three layers of iris-like rings made of highly reflective crystals (Gavelis et al., 2015; Hayakawa et al., 2015). Together, the lens and rings concentrate light onto the retinal body. The retinal body is separated from the lens by an invagination of the plasma membrane, forming an ocellar chamber filled with "vitreous" seawater. The dark red retinal body is concave and contains densely arranged stacks of waveform membranes that run in parallel to the incoming light (Greuet, 1968, 1977) (Figure 3.1). Despite this unusual arrangement, these membranes were recently shown to be modified thylakoids; in other words, the retinal body is a highly modified plastid (Gavelis et al., 2015). This reveals its homology to simper photoreceptive structures in other dinoflagellates, called "eyespots."

Eyespots are widespread among phototactic protists (Gavelis et al., 2017). They are either cuplike plastids or pigment clusters, sometimes with an overlying reflective layer. Eyespots serve to concentrate light on the base of the flagellum, where photoreceptor proteins are located. Pigmentation shields light from all but one direction—a necessity for directional phototaxis (Jekely, 2009). Unlike eyespots, however, the ocelloid possesses a lens, and it uses this to focus light onto the retinal body, rather than on the flagellum. Since the flagellum is located several microns away from the retinal body (in *Nematodinium*) or on the opposite end of the cell (in *Erythropsidinium*), it is unclear how the ocelloid transmits phototactic signals throughout the cell. In fact, there is still no direct evidence that the ocelloid can detect light at all. While the fragility of warnowiids has precluded behavioral observations, the vast size and high percentage of repetitive and non-coding sequences of dinoflagellate nuclear genomes currently makes warnowiids (and all but the smallest dinoflagellates) intractable for genomic surveys (Lin et al., 2015). A few warnowiid EST sequences are available, but the only sequenced photoreceptor protein is a xanthorhodopsin—a light-driven pump rather than a sensory-type rhodopsin (Hayakawa et al., 2015).

Nevertheless, work on wild-caught cells is allowing certain longstanding questions to be answered, such as the demonstration that the retinal body is a plastid. Moreover, *Erythropsidinium*—the genus with the largest ocelloid—has been reported to possess a striated fiber that could control ocelloid movement. While no evidence for this was provided at the time of description (Dodge and Greuet, 1987), new videographic data has confirmed that the ocelloid can indeed swivel 45°, suggesting that some cytoskeletal element functions akin to an oculomotor muscle (Gómez, 2017). Still, the question of the ocelloid's sensory capabilities remains unknown, as do basic details of warnowiid diversity, ecology, genetics, and behavior.

3.3 STATOCYST-LIKE ORGANELLES

As with photoreception, georeception (syn. gravireception)—the sensing of gravity—has led to the evolution of complex sensory structures across the tree of animals. The most widespread type of georeceptive organ is called a "statocyst," which consists of a fluid-filled spherical chamber. In its center is a dense mineralized "statolith" that can shift freely as the organism changes position relative to the gravitational field, thereby stimulating mechanoreceptor cells that line the chamber. Among animals, there are many variations on this theme, including statocysts lined by either multi-ciliated or uni-ciliated cells; statoliths that are either free floating or tethered to the statocyst wall; and statolith crystals made of either CaF_2, $CaSO_4$, $CaMgO_4P^+$, or $CaCO_3$ (Ariani et al., 1983; Wiederhold et al., 1989; Becker et al., 2005). Despite this variability, statoliths in some distant lineages of animals (e.g., cnidarians and snails) may be shaped by homologous genes, such as *Pax 2/5/8* and *POU-IV*, though statoliths in ctenophores exist in the apparent absence of *Pax* homologs (Kozmik et al., 2003; O'Brien and Degnan, 2003).

Animal statocysts generally consist of a dozen or more cells, but a case of subcellular convergent evolution has manifested in loxodid ciliates (Figure 3.2). These ciliates contain a spherical balancing organelle called the "Müller's vesicle," which contains a mineralized "Müller's body". The vacuole is about 7 μm in diameter and surrounded by flattened vesicles. The Müller's body is about 3 μm in diameter, bound by a double-membrane, and contains ~100 crystals of either $BaSO_4$ (in *Loxodes*) or $SrSO_4$ (in *Remanella*). This crystalized inclusion is suspended from the roof of the vesicle by a sheet of nine microtubules that connects to a modified cilium (immotile, but with a normal array of microtubules). In response to changes in the cell's position, the Müller's body's stalk can swing 90° as if on a hinge, articulating where the microtubules bind to a kinetosome (Fenchel and Finlay, 1986).

Depending on the species of loxodid ciliate, a cell can contain from one to thirty Müller's vesicles, often in different stages of morphogenesis. Development initiates with (1) formation of the crystalized Müller's body at the endoplasmic reticulum, followed by (2) migration of this body to an immotile cilium where (3) a vesicle envelops the crystalized body and cilium, (4) forming a functional Müller's vesicle containing the Müller's body connected by an internal ciliated stalk. Despite its subcellular origin, the Müller's vesicle has a similar overall structure and function to the statocysts in animals; changes in the position of the Müller's body prompt the ciliate to briefly "tumble" before altering course (Fenchel and Finlay, 1986). Moreover, laser ablation of the Müller's vesicles in *Loxodes* removed its geotactic behavior (Hemmersbach et al., 1998).

To our knowledge, no statocyst-like structures are found in other protists, even though many other lineages are geotactic. For instance, in *Euglena*, the entire cell is involved in georeception. Because the cell is denser than the surrounding freshwater environment, the weight of its cytoplasm creates tension on the underlying plasma membrane. Stretch-activated channels create a transmembrane calcium influx, informing which side of the cell is facing downward, even as it spins along its axis during helical swimming (Häder et al., 2017). Geotaxis can be disoriented in *Euglena* by placing cells in a medium with equal density, thereby relieving pressure on their mechanoreceptors. This treatment also disrupts geotaxis in *Paramecium* (which lacks Müller's vesicles), but it fails to disorient *Loxodes*, suggesting that its presently uncharacterized georeceptors are internal (Hemmersbach and Häder, 1999). Microgravity experiments

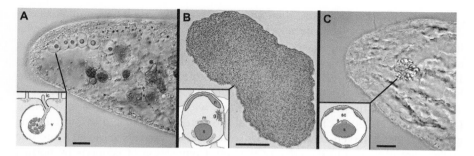

FIGURE 3.2 Statocysts and Müller's vesicle. (**A**) DIC light micrograph of the front of the ciliate *Loxodes magnus*, adorned by at least seven Müller's vesicles. Scale bar = 10 μm. **Inset**. Diagram of the Müller's vesicle, which consists of a statolith (s) suspected from an immotile cilium (ic), and is enclosed by a vacuole (v) lined by flattened vesicles called alveoli (a) (redrawn from Fenchel and Finlay, 1986). (**B**) Light micrograph of *Trichoplax adhaerens*, a deep-branching animal with a flat, nondescript body that lacks an antero-posterior axis but is fringed with up to 150 "crystal cells" around its perimeter (individual crystals not visible at this magnification). Scale bar = 100 μm. **Inset**. Diagram of a crystal cell in *Trichoplax*, showing that its Golgi apparatus (g) and cup-like nucleus (n) reside at the edge of the cell, while the central cytoplasm is occupied by a putative statocyst surrounded by mitochondria (m). (**C**) DIC light micrograph of an unidentified flatworm, in which the spherical crystal statolith has been shattered by pressure on the coverslip. Scale bar = 10 μm. **Inset**. Diagram of the statocyst of *Invenusta paracnida* (a rhabditophoran platyhelminth), in which the statocyst (s) is secreted by – and resides within – a lithocyte (lt) which can tumble freely within a statocyst chamber (sc) lined by sensory epithelium (Redrawn from Ehlers, 1991). Image A modified with permission from Michael Plewka (www.plingfactory.de). Image C modified with permission from Oliver Voigt.

have found georeception to be more sensitive in *Loxodes* than *Paramecium*, with detection limits of 0.16 g versus 0.3 g (Hemmersbach et al., 1996).

It is intriguing that loxodids have evolved statocyst-like organelles for geotaxis while other ciliates have not. This may have been selected for by the fine geotactic balance required by their environment and physiology. Loxodids dwell in marine sediments and stratified lakes and have low oxygen tolerances that restrict them to a narrow part of the oxycline. Even as the oxycline shifts due to tidal cycles or turbulence, loxodids maintain an optimal position via geotaxis, by swimming upward when O_2 levels are too low and downward when they are too high (Fenchel and Finlay, 1984). Given this acuity, why have no similar organelles been found among the myriad of other protists that navigate vertical gradients? Ciliates are larger than most other motile protists, and Fenchel and Finlay (1986) reasoned that the usefulness of statocyst-like vesicles in smaller organisms would be limited by size (Fenchel and Finlay, 1986). Even at a diameter of ~3 μm, (which is large by protist organelle standards) the Müller's body is visibly perturbed by Brownian motion, presumably contributing to sensory background noise. Given that smaller and less dense objects are more easily displaced, Fenchel and Finlay estimated that Müller's vesicles approached the lowest attainable size limits for functional statocyst-type organelles.

Surprisingly, this lower size limit was recently challenged by an animal, as evidenced by the ~2 μm geotactic crystals in *Trichoplax adhaerens*. As an "oligocellular" organism comprised of only six cell types, *Trichoplax* is the simplest free-living animal, with a flat, bilayer body fringed by "crystal cells" of previously unknown function. Each crystal cell is roughly spherical, bearing a central mass of rhomboid $CaCO_3$ crystals surrounded by mitochondria. A recent behavioral analysis of *Trichoplax* found these bodies to tumble freely within the cells, similar to statoliths (Mayorova et al., 2018). Their freedom of movement is facilitated by surrounding cytoplasm that is largely empty of other structures; endomembranes and a cup-like nucleus are displaced to the edge of the cell. Comparison of geotaxis by normal individuals (bearing up to 150 crystal cells) to that of rare individuals lacking crystal cells, found that only the former were able to regulate their depth under experimental conditions (Mayorova et al., 2018). In sum, statocysts seem to have manifested at three levels of organization across eukaryotes: subcellular (as in *Loxodes*), unicellular (as in *Trichoplax*), and multicellular (as in animals with statocysts)—with the first two lineages lacking nervous systems entirely. It remains to be seen whether the "statocyst" of *Trichoplax* is patterned by the same master regulatory genes as other animals. Interestingly, there is preliminary evidence that a *Pax 2/5/8* ortholog (*PaxB*) is expressed in the periphery of *Trichoplax*, where crystal cells are also found (Hadrys et al., 2005).

3.4 NEMATOCYST-LIKE ORGANELLES

Nematocysts (syn. "cnidocysts") are organelles responsible for the infamous stings of jellies, siphonophores, and other members of the Cnidaria and lend the group its name. Concentrated around the tentacles of the polyp or medusa, nematocysts function in prey capture and in defense (Stachowicz and Lindquist, 2000; Bullard and Hay, 2002). Each nematocyst acts as a "single-shot" ballistic weapon at the subcellular level. In one of the fastest mechanisms in biology, a capsule rapidly extrudes a hollow tubule—often in less than a microsecond—thereby puncturing and envenomating prey (Figure 3.3) (Holstein and Tardent, 1984; Nüchter et al., 2006). Discharge is driven by osmotic pressure up to 2,175 psi on the capsule wall, caused by an enrichment of poly-gamma-glutamate within the capsule lumen (Weber, 1989; Özbek et al., 2009). The capsule wall is comprised chiefly of minicollagens, cysteine-rich proteins with extensive disulfide-linkages that make it both tough and elastic (Pokidysheva et al., 2004; David et al., 2008). Upon discharge, the hollow tubule (nested within the capsule) turns inside out and uncoils. This rapid extension provides the force necessary to puncture prey, allowing the everted tubule to deliver its coating of paralytic toxins (Özbek et al., 2009). Most nematocysts are capped by a hatch-like operculum, and the presence/absence of other features, such as stylets and spines lining the tubule (Figure 3.3), have been used to classify nematocysts into several morphotypes, each specialized for certain prey items (Reft and Daly, 2012).

Nematocysts in cnidarians are but one type of projectile organelle, which exist in myriad forms across eukaryotes. Such organelles can be more broadly categorized as "extrusomes," membrane-bound compartments that discharge their contents outside

Convergent Evolution of Animal-Like Organelles 35

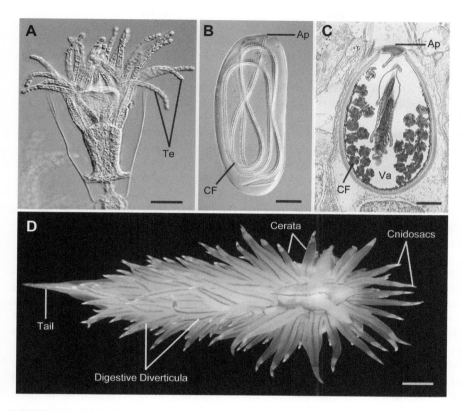

FIGURE 3.3 The position and structure of cnidarian nematocysts and their presence in some sea slugs via kleptocnidae (syn. nematoklepty). (**A**) Light micrograph of a gastrozooid in the hydroid *Obelia* showing the feeding tentacles equipped with cnidarian nematocysts within epidermal cells. (**B**) Light micrograph of a relatively large undischarged nematocyst from the moon jelly *Aurelia* showing the coiled filament that rapidly everts through the apical end of the capsule when stimulated. (**C**) Transmission electron micrograph through a cnidarian nematocyst showing multiple profiles through the coiled filament, the pressurized capsule chamber, and the position of the apical end. (**D**) A photograph of an aeolid nudibranch, *Hermissenda crassicornis*, showing the tentacle-like cerata on the dorsal surface, the position of the posterior end (tail), the digestive diverticula within each ceratum, and the cnidosacs at the tip of each ceratum. The cnidosacs are an extension of the gut lumen (i.e., digestive diverticula) and are filled with undischarged nematocysts stolen from cnidarian prey animals. The collection of (extracellular) nematocysts within the cnidosacs is discharged in a cloud of mucus through a pore when the nudibranch is threatened. Te = tentacles; CF = coiled filament; Ap = apical end; Va = pressurized vacuole. Scale bar = 100 μm in A; scale bar = 1 μm in B and C; scale bar = 0.2 μm in D. Image B is used with permission from W. Nell; image C used with permission from C. Moffet; Image D used with permission from M. LaBarbera.

the cell in response to either chemical or mechanical stimuli (Buonano and Ortenzi, 2018). Extrusome contents range from simple amorphous payloads to highly differentiated forms and can be involved in defense, predation, parasitism, cyst formation, and attachment to substrates (for the most thorough review of exrtrusome ultrastructure, see Hausmann, 1978). Although cnidarian nematocysts are the most intensively studied ballistic organelles, extrusomes are prevalent and diverse among protists, being found in many ciliates, dinoflagellates, euglenozoans, prasinophytes, cryptophytes, chrysophytes, raphidophytes, and chlorarachniophytes (Hausmann, 1978; Kugrens et al., 1994; Rosati and Modeo, 2003), as well as the rare, deep-branching eukaryotes *Ancoracysta* and *Hemimastix*. (Janouškovec et al., 2017; Lax et al., 2018). It is unclear how many times extrusomes evolved, given that their underlying genetics are poorly characterized and because it is difficult to infer homology between one ballistic mechanism and another based on morphology alone. Here, we discuss the "nematocysts" of dinoflagellates, which were the subject of one such homology-oriented debate and are remarkable from the standpoint of convergent evolution.

Characteristic of the dinoflagellate *Polykrikos*, these ~20 μm long extrusomes are only a fraction the size of cnidarian nematocysts but likewise possess a capsule, coiled tubule, stylet, and operculum (Gavelis et al., 2017) (Figure 3.4). Despite being named for their superficial similarity to nematocysts in cnidarians (Butschli, 1873), nematogenesis in *Polykrikos* is unique. Each nematocyst forms in five stacked vesicles that fuse at maturity, whereas cnidarians use a single nematogenic vesicle (Westfall et al., 1983). Furthermore, the tubule forms within the capsule, unlike in cnidarians, where the tubule develops as an outgrowth of the capsule wall, then inverts (Gavelis et al., 2017). Once assembled, the polykrikoid nematocyst migrates to the apical end of the cell and becomes positioned beneath a dense cylindrical organelle, the "taeniocyst," which is not found in cnidarians (Westfall et al., 1983; Hoppenrath et al., 2010) (Figure 3.2). Developmental differences aside, the polykrikoid nematocyst has a similar harpoon-like mechanism, in as far as the stylet is used to puncture prey, allowing the coiled tubule to inject into the prey cell (Chatton, 1914; Kofoid and Swezy, 1921).

This similarity inspired speculation that nematocysts in dinoflagellates and cnidarians are homologous, either via common ancestry or through the dissemination of certain components via horizontal gene transfer (Shostak, 1993; Hwang et al., 2008). However, recent comparative genomic surveys have failed to find evidence of cnidarian-type nematogenic genes (e.g. minicollagens, nematogalectin, or spinalin) in protists and found these genes to be restricted to cnidarians (including their parasitic offshoots, the myxozoans) (Holland et al., 2011; Shpirer et al., 2014; Lin et al., 2015; Gavelis et al., 2017). The only non-metazoan gene that cnidarians are known to employ in nematogenesis is *pgsAA*—which is responsible for biosynthesis of the osmotic propellant poly-gamma-glutamate—and is inferred to have been acquired via horizontal gene transfer from bacteria (Denker et al., 2008). *PgsAA* has not been found in extrusome-bearing protists, where extrusome propellants remain uncharacterized.

Investigations of polykrikoid feeding mechanisms have shown that the specifics of nematocyst function are quite divergent from cnidarians. Microscopic observations

Convergent Evolution of Animal-Like Organelles 37

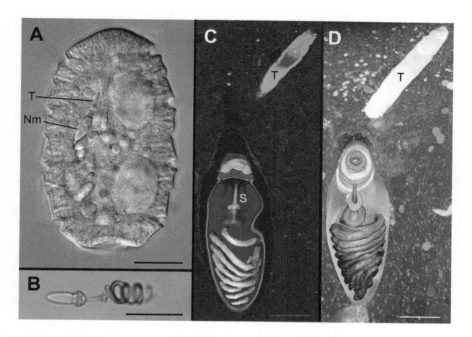

FIGURE 3.4 The structure of "nematocysts" in the dinoflagellate *Polykrikos kofoidii*. (**A**) DIC light micrograph of *P. kofoidii*. Polykrikoids possess two or more nuclei per cell, as well as taeniocysts (T) and nematocysts (Nm). (**B**) A discharged nematocyst in which the tubule and an everted portion of the capsule have been stained with Alcian blue. (**C-D**) Tridimensional reconstructions, based on focused ion beam scanning electron microscopy (FIB-SEM), of a nearly mature nematocyst and taeniocyst that have not yet migrated to the cell periphery. (**C**) Shows that the stylet (S) is poised to pierce the capsule from within. (**D**) Depicts the "operculum" as a complex nozzle formed of three concentric rings (yellow, orange, and red). Upon discharge, the stylet is rapidly ejected through the nozzle, followed by the tubule (purple), which uncoils. The stylet may also function to puncture prey. Scale bar = 20 µm in A & B, 2 µm in C & D.

suggest that the taeniocyst (a dense cylindrical structure positioned distally to the nematocyst in *Polykrikos*) is also an extrusome, as it is osmotically charged and becomes embedded in the prey cell upon contact (Gavelis et al., 2017). Feeding in *Polykrikos* involves two rapid steps: First, the prey contacts the taeniocyst and is adhered, then the nematocyst injects its coiled tubule into the prey cell. The mechanism by which the nematocyst discharges in polykrikoids differs from cnidarians in at least three ways (Gavelis et al., 2017); (1) Prior to exiting the capsule, both the stylet and tubule are launched through a concentric series of three rings, or "nozzle," that appears unique to dinoflagellates. (2) The stylet then punctures the capsule itself, which is entirely enclosed (whereas the capsule in cnidarians possesses a pre-existing opening through which it everts) and is only then able to contact the prey. (3) The tubule is mucilaginous and dissolves in seawater within a minute at room temperature, suggesting that—when injected into prey—it acts as a soluble delivery system. By contrast, the tubule in cnidarians is formed from insoluble minicollagens proteins, with toxins delivered by an overlying coating. Unfortunately, the contents of the tubule in polykrikoids—or any component of dinoflagellate nematocysts—have yet to be identified.

During this encounter, a motile prey cell may swim away or—if it is another dinoflagellate—may rapidly displace itself by discharging trichocysts. However, as long as the nematocyst is embedded in the prey cell, it remains tethered to *Polykrikos* by a "tow line." (The tow line is not to be confused with the tubule, as it does not originate from within the nematocyst capsule). By an unknown mechanism, *Polykrikos* can retract this tow line into the cytoplasm, drawing in the prey cell to be phagocytosed (Matsuoka et al., 2000). Thus, while the overall act of polykrikoid prey capture can be said to be "harpoonlike"—it includes several unique nematocyst features (a nozzle comprised of concentric rings, a tubule that is mucilaginous rather than insoluble, and a capsule that is completely sealed) and is accompanied by organelles with no clear analogs in cnidarians or other animals (the taeniocyst and the tow line).

As with any case of convergent evolution, the superficial similarities between nematocysts of cnidarians and dinoflagellates probably result from a limited morphospace for biological projectiles. Indeed, the eversible-tube-in-a-pressurized-capsule layout has been put to use in rapidly firing organs/organelles in diverse lineages and at many levels of organization. Multicellular versions of an eversible tube within a capsule are found in the proboscis of nemertean worms, cone snails, and platyhelminths (as the proboscis in kalyptorynchs and as the paracnids in coelogynoporids) and are usually propelled by muscular constriction of a hydrostatic sack (Sopott-Ehlers, 1981; Rundell and Leander, 2014). Single-celled versions of an eversible tube within a capsule are found in microbial eukaryotes like microsporidian fungi and the oomycete *Haptoglossa*, both of which have spores that launch nuclei into the host cytoplasm via an eversible tubule (Glocking and Beakes, 2000) (Figure 3.5). Yet while all predators must capture prey, only a small subset have evolved harpoonlike capture mechanisms. To better understand the selective pressure that drove convergence in these groups, we should consider what ecological commonalities they possess. The most obvious is that these organisms (fungus-like oomycetes, cnidarians, flatworms, and snails) are immotile or slow relative to their prey. Even though *Polykrikos* is a motile planktonic cell, it contends with dinoflagellate prey that have a rapid escape response of their own (trichocysts: spear-like extrusomes that can rapidly polymerize to push the cell several cell-lengths away from its predator). Thus, predatory ballistic structures can pre-emptively immobilize fast prey via adhesion, entanglement, and/or injection of paralytic agents. Pressurized ballistic mechanisms provide not only speed but the force necessary to puncture prey armor, and this feature would be advantageous for intracellular parasites that must penetrate a host (e.g. *Haptoglossa* and microsporidians) (Figure 3.5). By delivering both force and speed without the requirement of advanced neuromuscle systems, ballistic organelles have allowed slow, and/or morphologically streamlined organisms to exploit a broad range of taxa.

While polykrikoids are only known to use nematocysts for predation, cnidarians use nematocysts for both predation and defense, making them unpalatable to most

Convergent Evolution of Animal-Like Organelles 39

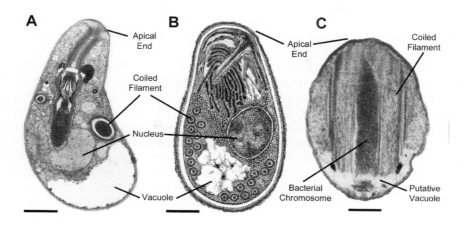

FIGURE 3.5 Convergent evolution of single-cells (in contrast to organelles) capable of discharging a coiled eversible filament within a capsule similar to cnidarian nematocysts. These three examples represent lineages that have diverged from each other over 1 billion years ago. (**A**) A transmission electron micrograph through the gun cell of *Haptoglossa*, a group of parasitic oomycetes (water molds) within the eukaryotic supergroup Stramenopila. (**B**) Schematic illustration of *Potaspora* showing the internal organization of a microsporidian spore, a parasitic relative of fungi within the eukaryotic supergroup Opisthokonta. Haptoglossan gun cells and microsporidan spores share a pressurized vacuole and coiled filament through which the nucleus moves into a host cell following discharge of the filament through the apical end. (**C**) A transmission electron micrograph through an epibiotic verrucomicrobial cell on the surface of the marine euglenozoan *Bihospites*. Like haptoglossans and microsporidians, these complex bacterial cells contain a putative vacuole and a tightly coiled filament that function to discharge a central bacterial chromosome through an apical pore. Scale bar = 0.5 μm in A; scale bar = 0.5 μm in B; scale bar = 0.2 μm in C. Image A modified with permission from Casal et al. (2008); image B modified with permission from Glocking and Beakes (2000); image C first described in Breglia et al. (2010).

non-specialist predators. Nematocysts are so effective at defending these soft-bodied invertebrates that several animal lineages have evolved to "steal" nematocysts for their own defense, with "nematoklepty" estimated to evolve from 9 to 17 times based on phylogenetic inference (Figure 3.3). This has originated most often among platyhelminths but is also found in aeolid nudibranchs and the ctenophore *Haeckelia*—all predators that digest cnidarian tissues and sequester the undischarged nematocysts within epithelial sacs, which they can eject when threatened (Mills and Miller, 1984; Vorobyeva et al., 2017; Krohne, 2018). Multiple lineages of crustaceans have also appropriated nematocysts indirectly through various forms of ectosymbiosis, such as by carrying cnidarians on mollusk shells, growing them directly on their carapaces, or—in the "boxer crab" *Lybia*—by wielding anemones in modified claws that they use both to ward off predators and to cleave the polyps for asexual propagation (Schnytzer et al., 2013, 2017). This diversity of "second-hand" extrusomes in mollusks, arthropods, platyhelminths, ctenophores, and even some protists (e.g., two genera of ciliates and euglenozoans host bacterial episymbionts that violently lyse upon contact to deter prey Breglia et al., 2010; Petroni et al., 2000; Rosati et al., 1999) (Figure 3.5)—suggests that strong selective pressures exist for the acquisition of ballistic defenses.

Nematogenic machinery, function, and development in cnidarians are highly conserved; all exhibit a basic harpoonlike function, varying only in size and the presence or absence of stylets and opercula. By contrast, not only do protists possess harpoon-like extrusomes, but also ballistic mechanisms involving proteinaceous syringes ("taeniocysts" and "toxicysts"), unfurling ribbons ("ejectisomes"), polymerizing spears ("trichocysts"), and lattices that rapidly expand upon hydration ("mucocysts") (Hausmann, 1978). Several predatory protists host multiple ballistic types simultaneously, with some specialized for predation and others for defense (e.g., a typical cell of *Polykrikos kofoidii* wields half a dozen nematocysts and taeniocysts, hundreds of mucocysts, and ~1,000 trichocysts), whereas cnidarians bear at most one nematocyst per cell. In protists, the molecular means by which extrusome arsenals are organized within the endomembrane system have only been experimentally studied in two ciliates (*Tetrahymena* bearing mucocysts, and *Paramecium* bearing trichocysts), and these studies implicate gene family expansions for CORVET and SNARE proteins, which facilitate vesicle tethering and fusion at membrane interfaces (Sparvoli et al., 2018; Plattner, 2017). It remains to be seen whether these same gene families expanded in dinoflagellates. While trichocyst matrix proteins appear homologous across dinoflagellates and ciliates based on sequence similarity (Rhiel, Wöhlbrand, and Rabus, 2018), proteins from other dinoflagellate extrusome types are as yet uncharacterized.

Even within a single extrusome type (nematocysts), dinoflagellates display greater structural variation than in cnidarians. For instance, not all dinoflagellate nematocysts have a clear harpoon-like arrangement. Nematocysts in warnowiids (aforementioned for their eye-like ocelloids) consist instead of 7 to 14 "barrels" in a radial arrangement reminiscent of a Gatling gun (Gavelis et al., 2017). Warnowiids are closely related to polykrikoids and share distinct features of nematocyst development (both involve multiple vesicles that fuse at maturity and become encased by a striated lattice with identical periodicity). The structural variation between nematocysts in these two groups is remarkable, given their relatively recent common ancestor (dated 120 MYA by molecular estimates), whereas cnidarians are likely to be at least four times as old, and their nematocysts show no deviations from a strict coiled-tubule-in-a-capsule formula (Cunningham et al., 2017; Žerdoner Čalasan, Kretschmann, and Gottschling, 2019). Genetic studies have found that the subtle differences across cnidarian nematocysts can be attributed in part to lineage-specific gene/domain duplications for proteins such as minicollagens (David et al., 2008; Shpirer et al., 2018), as well as the emergence of new protein isoforms (e.g., nematogalectin) via alternative splicing (Hwang et al., 2010). While the processes driving extrusome diversification in dinoflagellates are unknown, our understanding of them can be guided by the approaches used in the study of cnidarian nematogenesis. Given that *Polykrikos kofoidii* can be cultured, next steps should involve: (1) transcriptomics, (2) bioinformatic mining for potential structural proteins (e.g. cysteine-rich and coiled-coil domains are associated with the disulfide linkages and striated protein layers, which are common in extrusomes), (3) bulk proteomics on nematocysts to verify that candidate proteins are present, and lastly (4) immunolabeling to assess the location (and by inference, the function) of candidate proteins in charged and undischarged nematocysts.

3.5 CONCLUSIONS

Convergent evolution reflects similar selective pressures operating in similar environments. As discussed above, different examples of convergent evolution can occur across fundamentally different levels of biological organization and over vast phylogenetic distances. For instance, multicellular traits in some organisms have converged with either single-celled or subcellular traits in other organisms that have diverged from one another over 1 billion years ago. As in any comparative field, concepts such as "complexity" and "evolutionary lability" are relative. By incorporating protists in macroevolutionary discussions, we are able to consider animal cell features within a much larger morphospace. For instance, it is clear that protists have evolved certain sensory structures (e.g., statocyst-like Müller's vesicle and eye-like ocelloids) in the absence of gene complements that are familiar from animal models (e.g., *PaxB* and *Pax6* master control genes) and that they have formed these structures at fundamentally different levels of organization (organellar versus unicellular versus multicellular in animals). Not only do protists provide examples of extreme subcellular differentiation, they highlight how little we understand about the developmental processes that promote and constrain the evolution of complexity at the cellular scale.

REFERENCES

Ariani, A. P., Marmo, F., Balsamo, G., Franco, E., and Wittmann, K. J. 1983. The mineral composition of statoliths in relation to taxonomy and ecology in mysids. *Rapp. Comm. Int. l'Explor. Sci. Méditerr.* 28:333–336.

Arimoto, A., Nishitsuji, K., Higa, Y., Arakaki, N., Hisata, K., Shinzato, C., Satoh, N., and Shoguchi, E. 2019. A siphonous macroalgal genome suggests convergent functions of homeobox genes in algae and land plants. *DNA Res.* 26(2):183–192.

Barron, G. L. 1987. The gun cell of *Haptoglossa mirabilis*. *Mycologia* 79:877–883.

Becker, A., Sötje, I., Paulmann, C., Beckmann, F., Donath, T., Boese, R., Prymak, O., Tiemann, H., and Epple, M. 2005. Calcium sulfate hemihydrate is the inorganic mineral in statoliths of scyphozoan medusae (Cnidaria). *Dalt. Trans.* 8:1545–1550.

Breglia, S. A., Yubuki, N., Hoppenrath, M., and Leander, B. S. 2010. Ultrastructure and molecular phylogenetic position of a novel euglenozoan with extrusive episymbiotic bacteria: *Bihospites bacati* n. gen. et sp. (Symbiontida). *BMC Microbiol.* 10:145 (21 pages).

Bullard, S. G., and Hay, M. E. 2002. Palatability of marine macro-holoplankton: Nematocysts, nutritional quality, and chemistry as defenses against consumers. *Limnol. Oceanogr.* 47:1456–1467.

Buonano, F., and Ortenzi, C. 2018. Predatory-prey interactions in ciliated protists. In: *Extremophilic Microbes and Metabolites-Diversity, Bioprospecting and Biotechnological Applications*. London: InTechOpen, pp. 1–31.

Butschli, O. 1873. Einiges iiber Infusorien. Arch, mikrosk. *Anat. EntwMech.* 9:657–678.
Casal, G., Matos, E., Teles-Grilo, M. L., and Azevedo, C. 2008. A new microsporidian parasite, *Potaspora morhaphis* n. gen., n. sp. (Microsporidia) infecting the teleostean fish, *Potamorhaphis guianensis* from the River Amazon. Morphological, ultrastructural and molecular characterization. *Parasitology* 135:1053–1064.
Chatton, E. 1914. Les cnidocystes du peridinien: *Polykrikos schwartzi* Butschli. Albert Schulz. *Arch. Zool. Exp. Gén.* 54:157–194.
Cunningham, J. A., Liu, A. G., Bengtson, S., and Donoghue, P. C. 2017. The origin of animals: can molecular clocks and the fossil record be reconciled?. *BioEssays* 39(1):1–12.
David, C. N., Özbek, S., Adamczyk, P., Meier, S., Pauly, B., Chapman, J., Hwang, J. S., Gojobori, T., and Holstein, T. W. 2008. Evolution of complex structures: Minicollagens shape the cnidarian nematocyst. *Trends Genet.* 24:431–438.
Denker, E., Bapteste, E., Le Guyader, H., Manuel, M., and Rabet, N. 2008. Horizontal gene transfer and the evolution of cnidarian stinging cells. *Curr. Biol.* 18:R858–R859.
Dodge, J. D., and Greuet, C. 1987. Dinoflagellate ultrastructure and complex organelles. In: Taylor, F. J. R., editor. *The Biology of Dinoflagellates*. Oxford: Blackwell Publishing Inc., pp. 92–142.
Ehlers, U. 1991. Comparative morphology of statocysts in the Plathelminthes and the Xenoturbellida. In: *Turbellarian Biology*. Dordrecht: Springer, pp. 263–271.
Fenchel, T., and Finlay, B. J. 1984. Geotaxis in the ciliated protozoan Loxodes. *J. Exp. Biol.* 110:17–33.
Fenchel, T., and Finlay, B. J. 1986. The structure and function of Müller vesicles in loxodid ciliates. *J. Protozool.* 33:69–76.
Francis, D. 1967. On the eyespot of the dinoflagellate, *Nematodinium. J. Exp. Biol.* 47(3):495–501.
Funch, P., and Kristensen, R. M. 1995. Cycliophora is a new phylum with affinities to Entoprocta and Ectoprocta. *Nature* 378:711–714.
Gavelis, G., Hayakawa, S., White, III, R. W., Gojobori, T., Suttle, C., Keeling, P., and Leander, B. S. 2015. Eye-like ocelloids are built from different endosymbiotically acquired components. *Nature* 523:204–207.
Gavelis, G. S., Keeling, P. J., and Leander, B. S. 2017. How exaptations facilitated photosensory evolution: Seeing the light by accident. *BioEssays* 39:1600266.
Gavelis, G. S., Wakeman, K. C., Tillmann, U., Ripken, C., Mitarai, S., Herranz, M., Özbek, S., Holstein, T., Keeling, P. J., and Leander, B. S. 2017. Microbial arms race: Ballistic "nematocysts" in dinoflagellates represent a new extreme in organelle complexity. *Sci. Adv.* 3:e1602552.
Gehring, W. J., and Ikeo, K. 1999. Pax 6: Mastering eye morphogenesis and eye evolution. *Trends Genet.* 15:371–377.
Glocking, S. L., and Beakes, G. W. 2000. Two new species of *Haptoglossa*, *H. erumpens* and *H. dickii*, infecting nematodes in cow manure. *Mycol. Res.* 104:100–106.
Gómez, F. 2017. The function of the ocelloid and piston in the dinoflagellate *Erythropsidinium* (Gymnodiniales, Dinophyceae). *J. Phycol.* 53:629–641.
Goodheart, J. A., and Bely, A. E. 2017. Sequestration of nematocysts by divergent cnidarian predators: Mechanism, function, and evolution. *Invertebr. Biol.* 136:75–91.
Greuet, C. 1968. Organisation ultrastructurale de l'ocelle de deux peridiniens Warnowiidae, *Erythropsis pavillardi* Kofoid et Swezy et Warnowia pulchra Schiller. *Ed. Cent. Natl. Rech.* Sci. 4:209–230.
Greuet, C. 1977. Structural and ultrastructural evolution of ocelloid of *Erythropsidinium pavillardi* Kofoid and Swezy (dinoflagellate Warnowiidae, Lindemann) during division and palintomic divisions. *Protistologica* 13:127–143.
Häder, D. P., Braun, M., Grimm, D., and Hemmersbach, R. 2017. Gravireceptors in eukaryotes—A comparison of case studies on the cellular level. *Microgravity* 3:13.

Hadrys, T., DeSalle, R., Sagasser, S., Fischer, N., and Schierwater, B. 2005. The *Trichoplax PaxB* gene: A putative Proto-*PaxA/B/C* gene predating the origin of nerve and sensory cells. *Mol. Biol. Evol.* 22(7):1569–1578.

Hausmann, K. 1978. Extrusive organelles in protists. *Int. Rev. Cytol.* 52:197–276.

Hayakawa, S., Takaku, Y., Hwang, J. S. J., Horiguchi, T., Suga, H., Gehring, W., Ikeo, K., and Gojobori, T. 2015. Function and evolutionary origin of unicellular camera-type eye structure. *PLoS One* 10:e0118415.

Hemmersbach, R., Voormanns, R., Briegleb, W., Rieder, N., and Häder, D. P. 1996. Influence of accelerations on the spatial orientation of *Loxodes* and *Paramecium*. *J. Biotechnol.* 47:271–278.

Hemmersbach, R., Voormanns, R., Bromeis, B., Schmidt, N., Rabien, H., and Ivanova, K. 1998. Comparative studies of the graviresponses of *Paramecium* and *Loxodes*. *Adv. Sp. Res.* 21:1285–1289.

Holland, J. W., Okamura, B., Hartikainen, H., and Secombes, C. J. 2011. A novel minicollagen gene links cnidarians and myxozoans. *Proc. R. Soc. B Biol. Sci.* 278:546–553.

Holstein, T., and Tardent, P. 1984. An ultrahigh-speed analysis of exocytosis: Nematocyst discharge. *Science* 223:830–833.

Hoppenrath, M., Yubuki, N., Bachvaroff, T. R., and Leander, B. S. 2010. Re-classification of *Pheopolykrikos hartmannii* as *Polykrikos* (Dinophyceae) based partly on the ultrastructure of complex extrusomes. *Eur. J. Protistol.* 46:29–37.

Hwang, J. S., Nagai, S., Hayakawa, S., Takaku, Y., and Gojobori, T. 2008. The search for the origin of cnidarian nematocysts in dinoflagellates. In: *Evolutionary Biology from Concept to Application.* Berlin, Heidelberg: Springer, pp. 132–152.

Hwang, J. S., Takaku, Y., Momose, T., Adamczyk, P., Özbek, S., Ikeo, K., Khalturin, K., Hemmrich, G., Bosch, T. C., Holstein, T. W., and David, C. N. 2010. Nematogalectin, a nematocyst protein with GlyXY and galectin domains, demonstrates nematocyte-specific alternative splicing in *Hydra*. *PNAS* 107(43):18539–18544.

Janouškovec, J., Tikhonenkov, D. V., Burki, F., Howe, A. T., Rohwer, F. L., Mylnikov, A. P., and Keeling, P. J. 2017. A new lineage of eukaryotes illuminates early mitochondrial genome reduction. *Curr. Biol.* 27:3717–3724.

Jekely, G. 2009. Evolution of phototaxis. *Philos. Trans. R. Soc. B Biol. Sci.* 364:2795–2808.

Kofoid, C. A., and Swezy, O. 1921. The free-living unarmored Dinoflagellata. *Mem. Univ. Calif.* 5:1–538.

Kozmik, Z., Daube, M., Frei, E., Norman, B., Kos, L., Dishaw, L. J., Noll, M., and Piatigorsky, J. 2003. Role of pax genes in eye evolution: A cnidarian *PaxB* gene uniting *Pax2* and *Pax6* functions. *Dev. Cell* 5:773–785.

Krohne, G. 2018. Organelle survival in a foreign organism: *Hydra* nematocysts in the flatworm *Microstomum lineare*. *Eur. J. Cell Biol.* 97:289–299.

Kugrens, P., Lee, R. E., and Corliss, J. O. 1994. Ultrastructure, biogenesis, and functions of extrusive organelles in selected non-ciliate protists. *Protoplasma* 181:164–190.

Lax, G., Eglit, Y., Eme, L., Bertrand, E. M., Roger, A. J., and Simpson, A. G. B. 2018. Hemimastigophora is a novel supra-kingdom-level lineage of eukaryotes. *Nature* 564:410.

Leander, B. S., Saldarriaga, J. F., and Keeling, P. J. 2002. Surface morphology of the marine parasite *Haplozoon axiothellae* (Dinoflagellata). *Europ. J. Protistol.* 38:287–298.

Leander, B. S. 2008a. A hierarchical view of convergent evolution in microbial eukaryotes. *J. Eukaryot. Microbiol.* 55:59–68.

Leander, B. S. 2008b. Different modes of convergent evolution reflect phylogenetic distances: A reply to Arendt and Reznick. *Trends Ecol. Evol.* 23:481–482.

Lin, S., Cheng, S., Song, B., Zhong, X., Lin, X., Li, W., Li, L., Zhang, Y., Zhang, H., Ji, Z., et al. 2015. The *Symbiodinium kawagutii* genome illuminates dinoflagellate gene expression and coral symbiosis. *Science* 350:691–694.

Lukeš, J., Hashimi, H., and Zíková, A. 2005. Unexplained complexity of the mitochondrial genome and transcriptome in kinetoplastid flagellates. *Curr. Genet.* 48(5):277–299.

Lukes, J., Leander, B. S., and Keeling, P. J. 2009. Cascades of convergent evolution: The corresponding evolutionary histories of euglenozoans and dinoflagellates. *Proceedings of the national academy of sciences of the United States of America* 106 Supplement 1:9963–9970.

Marshall, W. F. 2011. Origins of cellular geometry. *BMC Biol.* 9(1):57.

Mast, F. D., Barlow, L. D., Rachubinski, R. A., and Dacks, J. B. 2014. Evolutionary mechanisms for establishing eukaryotic cellular complexity. *Trends Cell Biol.* 24(7):435–442.

Matsuoka, K., Cho, H.-J., and Jacobson, D. M. 2000. Observations of the feeding behavior and growth rates of the heterotrophic dinoflagellate *Polykrikos kofoidii* (Polykrikaceae, Dinophyceae). *Phycologia* 39:82–86.

Mayorova, T. D., Smith, C. L., Hammar, K., Winters, C. A., Pivovarova, N. B., Aronova, M. A., Leapman, R. D., and Reese, T. S. 2018. Cells containing aragonite crystals mediate responses to gravity in *Trichoplax adhaerens* (Placozoa), an animal lacking neurons and synapses. *PloS One* 13(1):e0190905.

Mills, C. E., and Miller, R. L. 1984. Ingestion of a medusa (*Aegina citrea*) by the nematocyst-containing ctenophore *Haeckelia rubra* (formerly *Euchlora rubra*): phylogenetic implications. *Mar. Biol.* 78:215–221.

Miyake, A., and Harumoto, T. 1996. Defensive function of trichocysts in *Paramecium* against the predatory ciliate *Monodinium balbiani*. *Eur. J. Protistol.* 32:128–133.

Mornin, L., and Francis, D. 1967. The fine structure of *Nematodinium armatum*, a naked dinoflagellate. *J. Microsc.* 6:759–780.

Nüchter, T., Benoit, M., Engel, U., Özbek, S., and Holstein, T. W. 2006. Nanosecond-scale kinetics of nematocyst discharge. *Curr. Biol.* 16:R316–R318.

O'Brien, E. K., and Degnan, B. M. 2003. Expression of *Pax2/5/8* in the gastropod statocyst: Insights into the antiquity of metazoan geosensory organs. *Evol. Dev.* 5:572–578.

Obst, M., Funch, P., and Kristensen, R. M. 2006. A new species of Cycliophora from the mouthparts of the American lobster, *Homarus americanus* (Nephropidae, Decapoda). *Organ. Div. Evol.* 6:83–97.

Özbek, S., Balasubramanian, P. G., and Holstein, T. W. 2009. Cnidocyst structure and the biomechanics of discharge. *Toxicon* 54:1038–1045.

Parfrey, L. W., Lahr, D. J., Knoll, A. H., and Katz, L. A. 2011. Estimating the timing of early eukaryotic diversification with multigene molecular clocks. *PNAS*, 108(33):13624–13629.

Parfrey, L. W., and Lahr, D. J. G. 2013. Multicellularity arose several times in the evolution of eukaryotes. *BioEssays* 35:339–347.

Petroni, G., Spring, S., Schleifer, K.-H., Verni, F., and Rosati, G. 2000. Defensive extrusive ectosymbionts of *Euplotidium* (Ciliophora) that contain microtubule-like structures are bacteria related to Verrucomicrobia. *PNAS* 97:1813–1817.

Plattner, H. 2017. Trichocysts—*Paramecium*'s projectile-like secretory organelles: Reappraisal of their biogenesis, composition, intracellular transport, and possible functions. *J. Eukaryot. Microbiol.* 64(1):106–133.

Pokidysheva, E., Milbradt, A. G., Meier, S., Renner, C., Häussinger, D., Bächinger, H. P., Moroder, L., Grzesiek, S., Holstein, T. W., Özbek, S., et al. 2004. The structure of the Cys-rich terminal domain of *Hydra* minicollagen, which is involved in disulfide networks of the nematocyst wall. *J. Biol. Chem.* 279:30395–30401.

Reft, A. J., and Daly, M. 2012. Morphology, distribution, and evolution of apical structure of nematocysts in hexacorallia. *J. Morphol.* 273:121–136.

Rhiel, E., Wöhlbrand, L., Rabus, R., and Voget, S. 2018. Candidates of trichocyst matrix proteins of the dinoflagellate Oxyrrhis marina. *Protoplasma* 255(1):217–230.

Rosati, G., Petroni, G., Quochi, S., Modeo, L., and Verni, F. 1999. Epixenosomes: Peculiar epibionts of the hypotrich ciliate *Euplotidium itoi* defend their host against predators. *J. Eukaryot. Microbiol.* 46:278–282.

Rosati, G., and Modeo, L. 2003. Extrusomes in ciliates: Diversification, distribution, and phylogenetic implications. *J. Eukaryot. Microbiol.* 50:383–402.

Rueckert, S., and Leander, B. S. 2008. Morphology and molecular phylogeny of *Haplozoon praxillellae* n. sp. (Dinoflagellata): A novel intestinal parasite of the maldanid polychaete *Praxillella pacifica* Berkeley. *Europ. J. Protistol.* 44:299–307.

Ruiz-Trillo, I., Burger, G., Holland, P. W. H., King, N., Lang, B. F., Roger, A. J., and Gray, M. W. 2007. The origins of multicellularity: A multi-taxon genome initiative. *Trends Genet.* 23:113–118.

Rundell, R. J., and Leander, B. S. 2010. Masters of miniaturization: Convergent evolution among interstitial eukaryotes. *BioEssays* 32:430–437.

Rundell, R. J., and Leander, B. S. 2014. Molecular examination of kalyptorhynch diversity (Platyhelminthes: Rhabdocoela), including descriptions of five meiofaunal species from the north-eastern Pacific Ocean. *J. Mar. Biol. Assoc. UK* 94:499–514.

Shpirer, E., Diamant, A., Cartwright, P., and Huchon, D. 2018. A genome wide survey reveals multiple nematocyst-specific genes in Myxozoa. *BMC Evol. Biol.* 18(1):138.

Schnytzer, Y., Giman, Y., Karplus, I., and Achituv, Y. 2013. Bonsai anemones: Growth suppression of sea anemones by their associated kleptoparasitic boxer crab. *J. Exp. Mar. Biol. Ecol.* 448:265–270.

Schnytzer, Y., Giman, Y., Karplus, I., and Achituv, Y. 2017. Boxer crabs induce asexual reproduction of their associated sea anemones by splitting and intraspecific theft. *PeerJ* 5:e2954.

Shostak, S. 1993. A symbiogenetic theory for the origins of cnidocysts in Cnidaria. *BioSystems* 29:49–58.

Shpirer, E., Chang, E. S., Diamant, A., Rubinstein, N., Cartwright, P., and Huchon, D. 2014. Diversity and evolution of myxozoan minicollagens and nematogalectins. *BMC Evol. Biol.* 205:1–12.

Sopott-Ehlers, B. 1981. Ultrastructural observations on paracnids. I: *Coelogynopora axi* Sopott (Turbellaria, Proseriata). In: *The Biology of the Turbellaria*. Dordrecht: Springer, pp. 253–257.

Sparvoli, D., Richardson, E., Osakada, H., Lan, X., Iwamoto, M., Bowman, G. R., Kontur, C., Bourland, W. A., Lynn, D. H., Pritchard, J. K., and Haraguchi, T. 2018. Remodeling the specificity of an endosomal CORVET tether underlies formation of regulated secretory vesicles in the ciliate Tetrahymena thermophila. *Curr. Biol.* 28(5):697–710.

Stachowicz, J. J., and Lindquist, N. 2000. Hydroid defenses against predators: The importance of secondary metabolites versus nematocysts. *Oecologia* 124:280–288.

Taylor, D. M., Lynn, D. H., and Gransden, S. G. 1995. *Vasichona opiliophila* n.sp., an ectosymbiotic ciliate (Chonotrichia; Exogemmida) on the maxillae of the snow crab, *Chionoecetes opilio*. *Can. J. Zool.* 73:166–172.

Taylor, F. J. R. 1980. On dinoflagellate evolution. *BioSystems* 13(1–2):65–108.

Vorobyeva, O. A., Ekimova, I. A., and Malakhov, V. V. 2017. The structure of cnidosacs in nudibranch mollusc *Aeolidia papillosa* (Linnaeus, 1761) and presumable mechanism of nematocysts release. *Dokl. Biol. Sci.* 476:196–199.

Wakeman, K. C., Yamaguchi, A., and Horiguchi, T. 2018. Molecular phylogeny and morphology of *Haplozoon ezoense* n. sp. (Dinophyceae): A parasitic dinoflagellate with ultrastructural evidence of remnant non-photosynthetic plastids. *Protist* 169:333–350.

Weber, J. 1989. Nematocysts (stinging capsules of Cnidaria) as Donnan-potential-dominated osmotic systems. *Eur. J. Biochem.* 184:465–476.

Westfall, J. A., Bradbury, P. C., and Townsend, J. W. 1983. Ultrastructure of the dinoflagellate *Polykrikos*. I. Development of the nematocyst-taeniocyst complex and morphology of the site for extrusion. *J. Cell Sci.* 63:245–261.

Wideman, J. G., Novick, A., Muñoz-Gómez, S. A., and Doolittle, W. F. 2019. Neutral evolution of cellular phenotypes. *Curr. Opin. Genet. Dev.* 58:87–94.

Wiederhold, M. L., Sheridan, C. E., and Smith, N. K. 1989. Function of molluscan statocysts. In: *Origin, Evolution, and Modern Aspects of Biomineralization in Plants and Algae.* Boston, MA: Springer, pp. 393–408.

Yubuki, N., and Leander, B. S. 2018. Diversity and evolutionary history of the Symbiontida (Euglenozoa). *Front. Ecol. Evol.* 6:100.

Žerdoner Čalasan, A., Kretschmann, J., and Gottschling, M. 2019. They are young, and they are many: Dating freshwater lineages in unicellular dinophytes. *Environ. Microbiol.* 21(11):4125–4135.

4 Evolution of the Animal Germline
Insights from Animal Lineages with Remarkable Regenerating Capabilities

Ana Riesgo and Jordi Solana

CONTENTS

4.1	Introduction	48
4.2	Features of Germ Cells	48
	4.2.1 Morphological Features	48
	4.2.2 Methods of Germline Determination	49
	4.2.3 The Conserved Genetic Machinery for Germline Determination	51
4.3	Porifera	52
	4.3.1 The Stem and Germ Cell Lines in Sponges: Archaeocytes and Choanocytes	52
	4.3.2 Morphological Features of Germ Cells in Sponges	56
	4.3.3 Molecular Signatures of Germ Cells in Sponges	56
	4.3.4 Determination of the Germline During Sponge Embryogenesis	57
4.4	Cnidaria	58
	4.4.1 The Origin of Gametes in Cnidarians: Ectoderm or Endoderm?	58
	4.4.2 Gene Expression of Germline Markers in Cnidarians and the Role of Nuage in the Specification of the Germline	59
4.5	Placozoa	60
4.6	Ctenophora	60
4.7	Acoela	61
4.8	Platyhelminthes	62
4.9	Concluding Remarks	63
Acknowledgements		64
References		65

4.1 INTRODUCTION

Germ cells carry hereditary information for the next generation (Wylie, 1999); hence, these cells must remain totipotent—they should retain the capacity to differentiate into each and every cell type of all the different organs (Raz, 2000)—and at the same time, they should differentiate into highly specialized gametes (Saffman and Lasko, 1999). The segregation of the germline from somatic cells during embryogenesis is considered essential for development and evolution (Wylie, 1999), but not all multicellular organisms segregate their germline during this process. In Volvocacea, segregation of somatic and germinal cells (gonidia) during embryogenesis occurs when the *glsA* gene determines the plane of cell division which results in cells with very different sizes; the large ones will become gonidia, and the rest will become somatic cells (Green and Kirk, 1981; Kirk et al., 1993; Miller and Kirk, 1999). Plants and animals undergo conventional embryogenesis and rely on the capacity of their gametes to create a whole new organism. However, only animals segregate their germline during the process of embryogenesis. In plants, there is no early deposition of germline since development is postembryonic, with the embryo bearing only primordial cells for apical growth (shoot and roots) (Steeves and Sussex, 1989). Plant gametes are formed once the organism has fully formed by differentiation from somatic cells through asymmetric divisions and differential expression of cis-regulatory elements (such as DUOI transcription factors), with extensive DNA methylation reprogramming (Takeda and Paszkowski, 2006; Peters et al., 2017).

Early germline segregation was originally thought to occur only in some animal groups (Gilbert ,2000), whereas in poriferans, cnidarians, flatworms, and tunicates, the premise was that somatic cells could readily become germ cells even in adult organisms. Today, this view is being largely abandoned, allowing a new approach that considers that at least in eumetazoans (i.e., Ctenophora, Cnidaria, and Bilateria) the germline is determined during embryonic development. Here, we review common aspects of the germ lineage among early-splitting animal groups (in terms of morphology, genetic machinery, and method of segregation), in order to seek common patterns that may be recognizable in early-splitting animals with remarkable regeneration capacities (see each specific section for details about regeneration), including sponges, cnidarians, ctenophores, placozoans, acoels, and ultimately platyhelminths (Figure 4.1), for these two last groups have remarkable similarities.

4.2 FEATURES OF GERM CELLS

4.2.1 Morphological Features

In many animals, germ cells are relatively easily distinguished from somatic cells by their histological characteristics. A widely conserved cytological characteristic of germ cells is the presence of the nuage (often referred to as germ granules) (Figure 4.2). This organelle is an RNA-rich structure that has been identified in many different animals (Eddy, 1975). The nuage is a discrete, dense, fibrous organelle, which is unbounded by a membrane, and is usually found among a mitochondrial cluster (Eddy, 1975; Saffman and Lasko, 1999), but often it is found around the

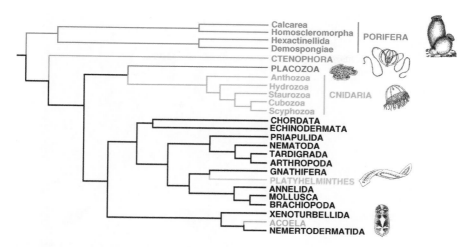

FIGURE 4.1 Schematic phylogenetic tree to show the relative position of the groups discussed here (Porifera, Ctenophora, Cnidaria, Placozoa, Platyhelminthes, and Acoela, in color). [Tree based on Laumer et al. (2018), Hejnol and Pang (2016), and Zapata et al. (2015).]

nucleus and close to nuclear pores (Seydoux and Braun, 2006). The principal components of the nuage are proteins and RNAs. Many of the genes that are implicated in germline specification have been found to be associated with the nuage or germ granules including genes such as *vasa*, *nanos*, and *tudor* (Seydoux and Braun, 2006).

4.2.2 Methods of Germline Determination

Germ cell specification occurs in different ways in animals, but two patterns of germ segregation are generally understood to occur in metazoans: **maternal determination and induction** (Buss, 1988; Ikenishi, 1998; Saffman and Lasko, 1999; Extavour and Akam, 2003; Seydoux and Braun, 2006). **Maternal determination** involves the synthesis of germ specific messenger RNAs and proteins during oogenesis and their transport to a specific location in the unfertilized egg or oocyte called the germ plasm (nuage). Cells inheriting this germ plasm become primordial germ cells (PGCs), and these will become the future germ cells. Maternal determination is clearly observed in *C. elegans*, *Drosophila*, and *Xenopus* (Ikenishi, 1998; Saffman and Lasko, 1999) as well as chaetognaths (Carré et al., 2002). In **induction**, determination of PGCs takes place much later during embryogenesis without any maternal inheritance of specific mRNAs by these germ cells. In this case, some cells are able to interpret the inductive signals arriving from neighboring tissues and express germline competence genes, which trigger them to differentiate into PGCs. This mechanism is suggested to occur in the cnidarians *Podocoryne* (Torras et al., 2004; Seipel et al., 2004), *Hydra* (Mochizuki et al., 2001), *Nematostella* (Extavour et al., 2005), and the mollusk *Ilyanassa* (Swartz et al., 2008) and was unequivocally proved in the mouse (McLaren, 2003). In other animals, PGCs are specified from embryonic or adult pluripotent cell populations. Frequently, these have nuage

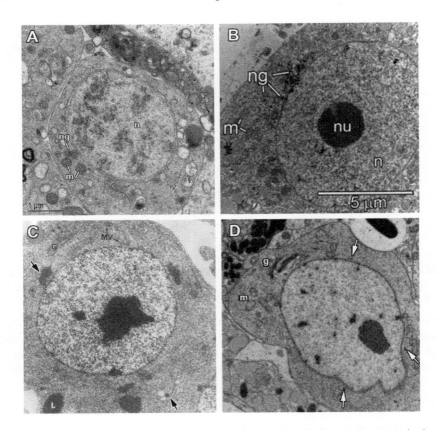

FIGURE 4.2 Nuage in poriferans, cnidarians, and planarians. A. Late spermatogonia showing nuage (ng) among mitochondria in the poriferan *Crambe crambe*. B. Oocyte showing nuage (ng) between a mitochondrial cluster in the poriferan *Paraleucilla magna*. C. Detail of an i-cell of the cnidarian *Pelmatohydra robusta*, showing nuage (arrows) in the vicinity of the nucleolated nucleus. D. Nuage (black arrowheads) located close to the nucleolated nucleus of a neoblast in the planarian *Dugesia tigrina*. Abbreviations: m and mv, mitochondria; g, Golgi apparatus; n, nucleus; nu, nucleolus. [Figures reproduced with permission from Lanna and Klautau (2010), Noda and Kanai (1977), and Auladell et al. (1993).]

particles, and often, these particles can be found already in the zygote—therefore, they are maternally deposited. Thus, in these animals, the germ cells are not segregated early, but a maternal determination event specifies the cell population that will ultimately produce the germ cells. This model has been called **intermediate** by Nieuwkoop and Sutasurya (1981) and is suggested to occur during the embryogenesis in the gastropod *Haliotis asinina* (Kranz et al., 2010), the polychaete *Platynereis dumerilii* (Rebscher et al., 2007), and the colonial ascidians *Botrylloides violaceus* (Brown and Swalla, 2007) and *Botryllus schlosseri* (Brown et al., 2009). In addition, some animals have populations of pluripotent stem cells in their adults that can regenerate the germline. Such is the case in planarians, for instance (Wang et al., 2007). This has led to the proposal that these populations, collectively known as **Primordial Stem Cells** (PriSCs), should be considered part of the germ line cycle (Solana, 2013). These populations—either embryonic or adult—can give rise to both somatic tissue and germ cells. In 1988, Buss suggested that when gametes are derived from a stem cell line that is mitotically active throughout ontogeny, even performing somatic functions, as occurs in sponges and cnidarians, it should be called multipotency. In contrast, Extavour and Akam (2003) suggest that the ancestral metazoan mechanism for germline specification was inductive and that germ plasm specification by maternal determination evolved as a derived character several times in different groups. But regardless of the mechanism of germ cell segregation, all animals seem to share the same genetic complement to build competent gametes, which is presented below.

4.2.3 THE CONSERVED GENETIC MACHINERY FOR GERMLINE DETERMINATION

Molecular studies have shown that embryonic or adult pluripotent populations that can give rise to somatic tissues (PriSCs) share molecular components with germ cells (Juliano et al., 2010; Juliano and Wessel, 2010; Alié et al., 2011), an issue that will be discussed later. Post-transcriptional regulation is key for the determination and maintenance of the germline (Seydoux and Braun, 2006). RNA regulators represent the core of the "germline program," controlling the processing, localization, translation, and degradation of RNA (Cinalli et al., 2008). Germ plasm components are typically RNA binding proteins. A number of genes are involved in germline specification (Cinalli et al., 2008; Seydoux and Braun, 2006; Strome and Updike, 2015; Lehman, 2016; Figure 4.3), collectively known as the germline multipotency program or GMP (Juliano et al., 2010). Among germline-specific genes, some have been also described as critical for germline specification, including *vasa*, *nanos*, and *piwi* (Figure 4.3). These genes are widely conserved throughout the Metazoa (e.g., Mochizuki et al., 2001; Extavour and Akam, 2003; Extavour et al., 2005; Sato et al., 2006; Rebscher et al., 2007; Grimson et al., 2008; Rabinowitz et al., 2008; Brown et al., 2009; Funayama, 2010; Kranz et al., 2010; see Figure 4.3). Interestingly, the protozoan *Giardia lamblia* possesses a *vasa*-like gene in its genome (Morrison et al., 2007), and a *vasa* locus (DDX4) has also been found in the choanoflagellate *Monosiga brevicollis* (Brown et al., 2008) although its true affiliation and significance still remains uncertain.

Vasa is a DEAD-box RNA helicase involved in the specification of the germ lineage. *Vasa* was first identified in *Drosophila* (Schüpbach and Wieschaus, 1986), and today, it is considered the most useful germ cell marker for Metazoa (Ewen-Campen et al., 2010; Gustafson and Wessel, 2010). *Nanos* is also involved in the germline specification (e.g., Johnstone and Lasko, 2001; Ewen-Campen et al., 2010) regulating genes to repress somatic identity (Hayashi et al., 2004) and preventing apoptosis (Hayashi et al., 2004; Sato et al., 2007). Piwi proteins have been found in eukaryotes (Cerutti et al., 2000) and are involved in germline determination in different animals (Cox et al., 2000; Kuramochi-Miyagawa et al., 2004; Girard et al., 2006; Megosh et al., 2006; Grimson et al., 2008), but they are also expressed in the stem

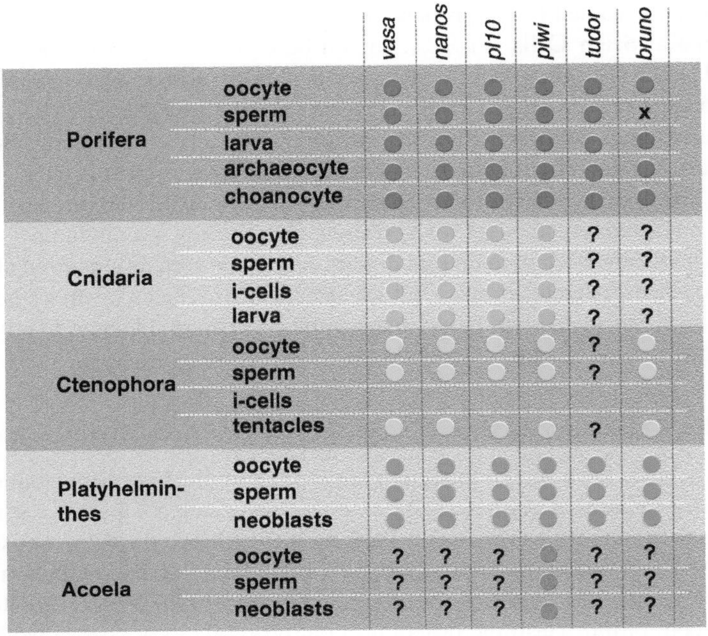

		vasa	nanos	pl10	piwi	tudor	bruno
Porifera	oocyte	●	●	●	●	●	●
	sperm	●	●	●	●	●	x
	larva	●	●	●	●	●	●
	archaeocyte	●	●	●	●	●	●
	choanocyte	●	●	●	●	●	●
Cnidaria	oocyte	●	●	●	●	?	?
	sperm	●	●	●	●	?	?
	i-cells	●	●	●	●	?	?
	larva	●	●	●	●	?	?
Ctenophora	oocyte	●	●	●	●	?	●
	sperm	●	●	●	●	?	●
	i-cells						
	tentacles	●	●	●	●	?	
Platyhelminthes	oocyte	●	●	●	●	●	●
	sperm	●	●	●	●	●	●
	neoblasts	●	●	●	●	●	●
Acoela	oocyte	?	?	?	●	?	?
	sperm	?	?	?	●	?	?
	neoblasts	?	?	?	●	?	?

FIGURE 4.3 Summary of expression of the most commonly reported genes of the germline multipotency program (GMP) in poriferans, cnidarians, ctenophores, planarians, and acoels.

cells (Carmell et al., 2002; King et al., 2001; Rossi et al., 2006; Seipel et al., 2004; Funayama, 2008, 2010).

The current state of knowledge of the molecular functions and biochemical connections between germline gene products has been reviewed by Ewen-Campen and collaborators (2010), with insights into nematode, insect, and vertebrate germline genetic programs. While we have a moderate knowledge of the gene pathways leading to the segregation of the germ line in the model systems *Caenorhabditis elegans* and *Drosophila melanogaster*, zebrafish, *Xenopus laevis*, and mouse (Ephrussi and Lehmann, 1992; Reinke et al., 2000; Saffman and Lasko, 1999; Styhler et al., 2002; Extavour and Akam, 2003; Ewen-Campen et al., 2010), the information regarding the mechanisms of germ cell formation in other metazoans are still far from being unraveled. In the next section, we will focus on the germ cell origin, features, and specification methods shown by early-branching invertebrates: sponges, placozoans, cnidarians, ctenophores, acoels, and planarians, with special emphasis, however, on sponges.

4.3 PORIFERA

4.3.1 The Stem and Germ Cell Lines in Sponges: Archaeocytes and Choanocytes

Sponges show an impressive regenerative capacity, being able to reorganize their body after complete cell dissociation (although not all, see Eerkes-Medrano et al., 2015;

Lavrov and Kosevich, 2016). Such an astonishing capacity suggests a strong stem cell system that is capable of complete re-organization of the body from the few cells that survive dissociation (Eerkes-Medrano et al., 2015). In fact, sponges possess a very simple body plan with epithelial layers and a mesohyl containing relatively few types (Simpson, 1984). Among the epithelial cells, the choanocyte is a specialized flagellated cell that captures food particles using the water flow movement. Choanocytes are considered to be pluripotent stem cells of sponges (Funayama, 2008, 2010, 2018; Funayama et al., 2010) since they are capable of trans-differentiating into several cell types, including archaeocytes (Connes et al., 1974). By contrast, archaeocytes are capable of differentiation into any other cell of the sponge. Therefore, archeocytes are likely totipotent stem cells but also have somatic metabolic functions (nutrient transport and digestion, see (Simpson, 1984; Müller, 2006; and Funayama, 2008, 2010, 2018 for reviews). Archeocytes are one of the most prominent mesohyl cell types (also referred as amoebocytes, polyblasts, or thesocytes by many authors, see Simpson [1984]) and have strikingly similar morphological features with stem cells (see Figure 4.4) such as a high nuclear/cytoplasmic ratio (Figure 4.4A–F) and a large nucleolate nucleus with a single nucleolus (Figure 4.4A–F). Archaeocytes can also show a compact basophilic cytoplasm, well-developed endoplasmic reticulum, and Golgi apparatus (Figure 4.4B–C, E), numerous ribosomes and mitochondria (Figure 4.4B–C, E), and phagosomes, lysosomes, and other vesicles (Isaeva et al., 2009; see Figure 4.2), but there is in fact a wide morphological variability among sponge archaeocytes, even within the same species (Figure 4.4). Some authors attribute the variability to different physiological states of the same cell (Funayama, 2018), but Müller (2006) already suggested that archaeocytes might not be a single population of cells but several that can only be identified with the use of molecular markers (e.g., Krasko et al., 2000; Perović-Ottstadt et al., 2004). Interestingly, although choanocytes have also been thought to be part of the sponge stem cell system (Funayama, 2008, 2010, 2018), their morphological traits have very little in common with stem cells (Figure 4.5), except for some choanocytes in the chambers of the homoscleromorph *Corticium candelabrum* that show a nucleolate nucleus (Figure 4.5D) when choanocytes usually never show a nucleolus (Simpson, 1984).

Sponges can reproduce both asexually (by gemmulation, budding, or fragmentation) and sexually (by either hermaphroditism or gonochorism), and their gametogenesis and morphology of gametes show a remarkable similarity to those of the rest of animals (Riesgo and Maldonado, 2009; Ereskovsky, 2010). Different cell types in the adult sponge have been pointed out as the origin of gametes, i.e., germ cells. In the early 20th century (reviewed in Hegner, 1914), early investigators considered germ cells as mesodermal in origin. At that time, they assumed that the sponge mesohyl, where archaeocytes are located, was equivalent to mesoderm. According to Weltner (1907), both amoebocytes and tokocytes (i.e., gonocyte precursors) are physiological states of the same cell, the archaeocyte. This would then imply that archaeocytes are the primordial germ cells of sponges, a view also shared by Schulze (1875). But choanocytes are also capable of becoming germ cells (see Simpson, 1984 and Ereskovsky, 2010 for reviews). As a general rule, archaeocytes usually give rise to oocytes, while choanocytes give rise to sperm. However, there are some exceptions to these rules. Sponge sperm derives from choanocytes in almost all sponges,

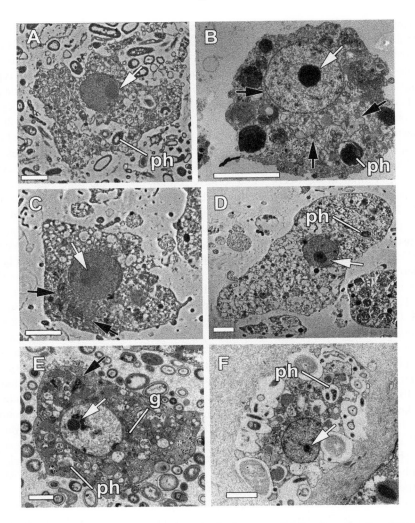

FIGURE 4.4 Archaeocytes in sponges. A. Archaeocyte in *Geodia atlantica*. Note the eccentrically nucleolated nucleus (white arrow) and the phagosomes (ph) containing engulfed bacteria. B. *Isodictya kerguelenensis*. The nucleolus is centered in the nucleus (white arrow). Note the abundant endoplasmic reticulum (black arrows) and the phagosomes (ph). C-D. Two morphologically different archaeocytes in *Halichondria panicea*. Note the centrically nucleolated nucleus (white arrows), the multiple endoplasmic reticuli (black arrows) and the phagosomes (ph). E. Archaeocyte in *Ircinia fasciculate* showing eccentrically nucleolated nucleus (white arrow), endoplasmic reticulum (black arrow), Golgi apparatus (g) and phagosomes (ph). F. Archaeocyte-like cell in *Corticium candelabrum* showing nucleolated nucleus (white arrow) and phagosomes (ph) containing bacteria. Scale bar: 2 μm.

Evolution of the Animal Germline

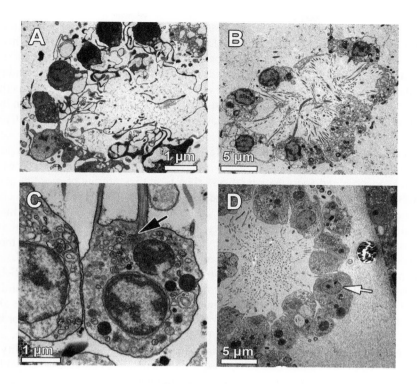

FIGURE 4.5 Choanocytes in sponges. Note their anucleolate nature, with diffuse and unorganized chromatin A. Choanocyte chamber in *Mycale acerata*. B. Choanocyte chamber in *Raspaciona aculeata*. C. Choanocyte of *Petrosia ficiformis* showing endoplasmic reticulum (black arrow). D. Choanocyte chamber in *Corticium candelabrum*, note the single choanocyte with a binucleolated nucleus (white arrow).

except for hexactinellids, carnivorous sponges, and few other exceptions, where choanocytes are absent or heavily modified (Leys et al., 2006; Riesgo et al., 2007a). While archaeocytes become oocytes in most demosponges and hexactinellids (see Simpson, 1984; Maldonado and Riesgo, 2008; and Ereskovsky, 2010 for reviews), and some homoscleromorphs (Riesgo et al., 2007b), studying calcareous sponges, where archaeocytes are not frequent, Haeckel (1872) reported that eggs were derived from the flagellated cells (choanocytes). Similarly, in some homoscleromorphs, oocytes derive from choanocytes (Gaino et al., 1986). Such apparent incongruence occurs in some other Demospongiae and Calcarea. But, as pointed out by Simpson (1984), histological observations are not proof enough of a differentiation of a cell into oogonia or spermatogonia, and therefore the accuracy of the interpretations should be taken with much caution, as was already pointed out in the early 20th century by Minchin (1900).

In the adult sponge, germ cells are not identifiable, and their study is still difficult due to several factors, including: 1. the absence of sexually reproducing sponge cultures, 2. the impossibility of long cell tracking during larval settlement until fully functional reproductive sponges, 3. the relatively limited molecular and genetic techniques available for the study of sponge development (Degnan et al., 2015). That is why today, for most developmental biologists, sponges are considered to lack any evident germ cell line (Gilbert, 2000), and therefore, its segregation during embryonic development is assumed as totally absent. In turn, the current hypothesis is that somatic adult cells (any) receive external stimuli that trigger their differentiation into germ cells at any point in their life. However, this point of view is in fact relatively recent.

According to the morphological studies of germ cells in sponges (see previous section), both choanocytes and archaeocytes are capable of becoming both types of gametes (oogonia and spermatogonia), and therefore, they are the primordial germ cells. Sarà (1974) and later Funayama and collaborators (2010), suggested the choanocyte as the gamete anlagen, although Sarà considered the possibility of an intermediate step as an archaeocyte. However, based on the results of transcriptomic analysis of *Ephydatia fluviatilis*, archaeocytes showed a transcriptional profile more consistent with that of stem cells (Alié et al., 2015). In the only single-cell RNAseq study performed in sponges (in the demosponge *Amphimedon queenslandica*) Sebé-Pedrós and collaborators (2018a) also showed a transcriptional program containing many GMP elements in archaeocytes, although a subpopulation of choanocytes also showed expression of *bruno* and *nanos*. We believe that the ancestral primordial germ cell that gave rise to both gamete types in sponges is the archaeocyte, and later on, during the diversification of sponges, they must have acquired the ability to transform archaeocytes into one gamete type and choanocytes into the other, with a potential intermediate archaeocyte step in this latter case.

4.3.2 Morphological Features of Germ Cells in Sponges

Animal germ cells can be readily identified by the presence of the nuage (Eddy, 1975), but it is only very recently that the nuage has also been identified in sponge cells: oocytes (Lanna and Klautau, 2010; Gonobovleva and Efremova, 2017; Figure 4.2B), late spermatogonia (Fierro-Constaín et al., 2017; Riesgo and Maldonado, unpublished; Figure 4.2A), and more interestingly, archaeocytes (Akhmadieva, 2008). Such evidence confirms the possibility of archaeocytes being the germline precursors. Some questions inevitably arise: do all archaeocytes contain germinal granules or only a small subpopulation? Are the germinal granules present throughout the life of the sponge or only when it is reproductive? Do choanocytes contain germ plasm as well?

4.3.3 Molecular Signatures of Germ Cells in Sponges

Besides the presence of the nuage, germ cells can be identified as well by their transcriptional activity, in particular by the expression of what are called germline markers or the germline multipotency program (GMP; Figure 4.3). This toolkit involves several genes (*vasa, piwi, nanos, PL10, boule, tudor, bruno, and pumilio*) that have been found in all sponge groups (Mochizuki et al., 2001; Grimson et al., 2008;

Funayama et al., 2010; Kerner et al., 2011; Leininger et al., 2014; Riesgo et al., 2014; Alié et al., 2015; Fierro-Constaín et al., 2017), but their expression patterns have been reported for very few sponge species. The germline gene *piwi* (two homologs, *EfpiwiA* and *EfpiwiB*) was found to be expressed mostly in choanocytes but also in archaeocytes of the freshwater sponge *Ephydatia fluviatilis* (Funayama et al., 2010). The first transcriptomic analysis, performed in *Ephydatia fluviatilis*, showed that archaeocytes differentially express the genes *vasa*, *Bruno*, and *musashi* but also *EfpiwiA* and *EfpiwiB* (Alié et al., 2015). In the same study, choanocytes were not observed to express any GMP marker (Alié et al., 2015), in contrast to previous studies (Funayama et al., 2010). In *Sycon ciliatum*, *PL10*, *vasa*, and *nanos* were expressed in oocytes and *PL10B* and *vasaB* in choanocytes (Leininger et al., 2014). Surprisingly, although *Scinanos* was found strongly expressed in the macromeres and cross cells during larval pre-inversion, their expression was absent in adult cells (Leininger et al. 2014). In non-reproducing *Oscarella lobularis*, certain subpopulations of choanocytes and archaeocytes (as well as pinacocytes when regeneration was taking place) express some of the germline markers (Fierro-Constaín et al., 2017). In *O. lobularis*, both gamete types were found to express all germline proteins (Fierro-Constaín et al., 2017). Interestingly, back in the 19th century, Fiedler (1888) observed that in *Spongilla* only certain amoeboid cells (archaeocytes) of the mesohyl became germ cells, which has been now supported by molecular studies.

But interestingly, the fact that not all archaeocytes and choanocytes express the full complement of GMPs in the sponges studied (see for instance Fierro-Constaín et al., 2017 and Sebé-Pedrós et al., 2018a) and that all GMP genes were also expressed in pinacocytes (epithelial cells) during wound healing in *O. lobularis* (Fierro-Constaín et al., 2017) suggests there are other roles for the GMP in somatic cells.

4.3.4 Determination of the Germline During Sponge Embryogenesis

In animals, cells in the early embryo undergo a series of morphogenetic movements or cellular reorganization resulting in ectoderm, endoderm, or mesoderm, and cell type specification often occurs during this period (Buss, 1983). During sponge embryogenesis, differentiated adult cells are observable in only a handful of sponges (e.g., Uriz et al., 2001; Maldonado et al., 2002; Leys et al., 2006; Riesgo et al., 2007a; Degnan et al., 2015). However, in *Amphimedon queenslandica*, apparently there is no constancy in larval and juvenile cell types; although there is cell differentiation during the embryogenesis, the final fate of the cell types is not determined during that period (Degnan et al., 2015). Whether that is also applicable to the germline still remains unanswered. In the 19th century, sponge researchers had already suggested that the germline was determined during embryogenesis by observing larval and juvenile morphogenetic processes. Maas (1893) distinguished two types of amoeboid cells in the mesohyl of the adult sponges, and these two cell types were also obvious in the larva, which led him to suggest that one of the cell types was the primordial germ cell type. Minchin (1900), and later Borojevic (1969), also noted the presence of undifferentiated blastomeres in the posterior pole of the larva of *Clathrina blanca* and hypothesized that those would give rise to both archaeocytes and tokocytes or gonocytes (germ cells) after metamorphosis.

Besides the suggestions of the sponge researchers in the 19th and early 20th century, we have very little evidence of the germline being segregated during embryogenesis in sponges. In *Amphimedon queenslandica*, *vasa*, *nanos*, and *PL10* are expressed in the micromeres of the larva (Degnan et al., 2015). In *Sycon ciliatum*, the two homologs of vasa were expressed, first broadly in the embryos and later upregulated in the cruciform cells of the larva (Leininger et al., 2014). All GMP genes were expressed in the blastula and larvae stages of *Oscarella lobularis* (Fierro-Constaín et al., 2017). But even though these are exciting discoveries, whether the larval cells expressing the GMPs are the ones that will give rise to the subpopulation of stem/germ cells that will become gametes still needs confirmation.

4.4 CNIDARIA

Cnidarians are also capable of massive body regeneration, including regeneration of amputated body or foot, budding, complete individual reassembly from dissociated cells, and grafting (Galliot and Schmid, 2003; Holstein et al., 2003). Such regeneration abilities are best known from hydrozoans (especially *Hydra*), which are relatively easily cultured and amenable for manipulation (Galliot and Schmid, 2003; Holstein et al., 2003). The study of germline determination and regeneration abilities of cnidarians have come hand in hand very often, and probably because of this, the germline determination of the class Hydrozoa (Figure 4.1) is the best studied and understood among the Cnidaria, although strong effort has been made recently to unravel the developmental patterns in some anthozoans (e.g., Finnerty and Martindale, 1999; Martindale et al., 2004; Extavour et al., 2005).

4.4.1 THE ORIGIN OF GAMETES IN CNIDARIANS: ECTODERM OR ENDODERM?

The embryonic origin for germ cells of anthozoans, cubozoans, hydrozoans, and scyphozoans is still a matter of debate. Interestingly, germ cells are known to originate in the gastrodermis of the mesentery and then move into the mesoglea where they mature in both anthozoans and scyphozoans (reviewed in Fautin and Mariscal, 1991; Lesh-Laurie and Suchy, 1991). In hydrozoans, the origin of germ cells has been a matter of discussion since the end of the 19th century. While some researchers thought that ova originate from endodermal cells and sperm from ectodermal cells (reviewed by Hegner, 1914), Schulze (1871) and Kleinenberg (1872) suggested that germ cells are derived from ectodermal cells, in particular the interstitial cells (i.e., I-cells) in *Hydra*. At some point in the early 20th century, the agreement was that germ cells were not derived from the ectoderm or the endoderm but belonged to a special sort of propagative cell scattered among the rest of cells and that they would give rise to ova or spermatozoa under certain environmental conditions (Hegner, 1914). But the following years witnessed again the controversy about the ectoderm/endoderm origin of germ cells (Brien and Reniers-Decoen, 1955; Noda and Kanai, 1977; Sugiyama and Sugimoto, 1985; Bosch and David, 1987; Nishimiya-Fujisawa and Sugiyama, 1993; Littlefield, 1994). Nowadays, it is generally accepted that a clear division between somatic cells and germline cells is not established in cnidarians,

Evolution of the Animal Germline

and germ stem cells arise de novo from I-cells (Seipel et al., 2004). Therefore, I-cells (or at least a small subpopulation of them) are currently the candidates for the origin of germ cells in cnidarians.

4.4.2 GENE EXPRESSION OF GERMLINE MARKERS IN CNIDARIANS AND THE ROLE OF NUAGE IN THE SPECIFICATION OF THE GERMLINE

Interstitial cells show similar features to primordial germ cells in other animals: undifferentiated round nucleolated cells, with affinity for basic dyes, a cytoplasm filled with ribosomes, and the presence of nuage (Noda and Kanai, 1977; Frank et al., 2009). A nuage also appears both in oocytes (e.g., Noda and Kanai, 1977; Eckelbarger and Larson, 1992) and sperm (Larkman, 1984) in many cnidarians. Nishimiya-Fujisawa and Kobayashi (2012) and Siebert and Juliano (2017) reviewed the evidence for germline specification in *Hydra* and other cnidarians, where a population of multipotent I-cells gives rise to self-renewing germline stem cells, which are then responsible for gamete production. However, the majority of the evidence available comes from experimental manipulation of cell lineages, as in Littlefield (1991), and therefore the natural commitment of I-cells to the germline is still unknown.

Mochizuki et al. (2000, 2001) performed the first study regarding cnidarian germline markers. They found expression of the germ cell markers belonging to the *nanos* (*Cnnos1*, *Cnnos2*) and *vasa* (*Cnvas1* and *Cnvas 2*) families, in the interstitial cells committed to germ cells in *Hydra*. The expression of *Cnnos1* and both *vasa* genes was restricted to multipotent stem cells (I-cells) and oocytes, being null in somatic cells (Mochizuki et al., 2000, 2001). Rebscher et al. (2008) studied the localization of the gene *vasa* (*HeVas*) in *Hydractinia echinata*, finding that *HeVas* was upregulated during gametogenesis in I-cells, but could not find mRNA transcripts on the unfertilized oocyte or during embryogenesis until much later in development. However, using an antibody against the vasa protein of *Platynereis dumerilii*, the authors found that the protein accumulated in the nuage of the oocytes during gametogenesis. In *Podocoryne carnea,* two *nanos* genes were found (*Pcnos1* and *Pcnos2*), and both were expressed in the fertilized egg and undifferentiated mitotically active cells of the early medusae, as well as in the gonads of mature medusae (Torras et al., 2004). In the anthozoan *Nematostella vectensis*, Extavour et al. (2005) found two *vasa* genes (*Nvvas1* and *Nvvas2*) and also two nanos genes (*Nvnos1* and *Nvnos2*), suggesting that the duplication of those genes predated the branching of cnidarians. While *Nvas1* was inherited as a maternal transcript in the oocyte, the expression of *Nvvas2* was not initiated until gastrulation. In *N. vectensis* as in *H. echinata*, Vasa protein was also accumulated during oogenesis in the oocyte (Figure 4.3). In siphonophores, particularly in *Nanomia bijuga*, it was only recently that I-cells were identified in growth zones using germline markers (*vasa-1*, *PL10*, *piwi*, *nanos-1*, and *nanos-2*), which were also expressed in gametogenic regions, indicating the origin of germ cells in I-cells (Siebert et al., 2015).

Interestingly, since in cnidarians the GMP markers are shown to be present in the multipotent cell lineage that gives rise to not only the gametes (i.e., I-cells), there is an emerging debate now of whether these genes drive the specification of germ cells or instead are involved in the maintenance of their toti-/multi-/pluri-potency program (Rebscher et al., 2008; Juliano et al., 2010).

4.5 PLACOZOA

The position of placozoans within the animal tree of life (Figure 4.1) is still contentious. While some authors consider the phylum as a sister group to other metazoans (i.e., Syed and Schierwater, 2002a,b; Schierwater et al., 2009a,b), others place them as derived cnidarians (Bridge et al., 1995; Sidall et al., 1995), sister to cnidarians (Laumer et al., 2018), or sister to the rest of Eumetazoa (including cnidarians) based on whole genome analysis (Srivastava et al., 2008). But many other positions for placozoans have been suggested within the base of the animal tree (reviewed in Srivastava et al., 2008). Placozoans are morphologically simple, but this contrasts with the complexity of their genome (Srivastava et al., 2008). Placozoans consist of only four different cell types: dorsal epithelial cells, gland and cylinder cells of the ventral epithelium, and fiber cells of the interspace (Grell and Ruthman, 1991). There are only two records of oocytes and embryos (Grell and Ruthman, 1991; Eitel et al., 2011) and sperm (Grell and Benwitz, 1981). Although the observations on sperm were only tentative, the putative eggs cleaved to yield up to 256 cells but failed to develop further (Srivastava et al., 2008). Also, population genetic analyses reveal allelic variation and evidence for genetic recombination in animals in the wild that is consistent with sexual reproduction (Signorovitch et al., 2005).

Other evidence for sex comes from the genome (Srivastava et al., 2008; Eitel et al., 2011). *Trichoplax* has meiosis-associated genes like meig1 and nme5 (Eitel et al., 2011) and germline markers including an ortholog of the zinc finger gene *nanos* and a member of the *vasa/PL10* family of DEAD box helicases, as well as homologs of *mago nashi*, *Par-1*, *pumilio*, and *tudor* (Srivastava et al., 2008), all implicated in primary germ cell development in eumetazoans, and sperm-associated protein genes (Eitel et al., 2011). In this sense, the challenge here relies on verifying germline formation in placozoans documenting the expression and functions of these GMP genes.

4.6 CTENOPHORA

The phylogenetic position of ctenophores is currently very controversial (e.g., Dunn et al., 2008; Hejnol et al., 2009; Pick et al., 2010; Ryan et al., 2013, Figure 4.1), which actually revived the interest for all aspects of their biology and evolution (e.g., Ryan et al., 2010; Moroz, 2015). Ctenophores have been the focus of regeneration research as well, since adults display remarkable abilities. All body parts are capable of regeneration, and in fact, if embryos are bisected, depending on the timing, they can regenerate the whole individual (Pang and Martindale, 2008). Interestingly, whereas individuals cut in halves, thirds, and quarters along the oral-aboral axis are capable of regeneration, those cut in eighths are not (see Pang and Martindale, 2008). Regarding germline specification, both Nieuwkoop and Sutasurya (1981) and Buss (1983) claimed a maternal determination mode of germ cell specification in this group. In contrast, Extavour and Akam (2003) believed that in ctenophores, the germ cells arise by induction, from the meridional canal endoderm, although they also considered that their migration from the meridional canal primordium cannot be ruled out. The embryonic origin of germ cells was impossible to find in the ctenophore *Mnemiopsis leidyi* (Martindale and Henry, 1999) using modern techniques of fate mapping because embryos did not remain labeled for long enough.

But the development of more sophisticated molecular techniques has helped fill the gap of knowledge about germline specification in ctenophores. Recently, Alié et al. (2011) found that *vasa*, *piwi*, *bruno*, and *PL10*, were expressed in the female and male germlines, and also (except for one *piwi* paralog) they showed similar expression patterns in stem cells within somatic territories. The authors suggest that this germline machinery is acting in two different contexts, the germline and the stem cells, suggesting a role of these genes mainly in maintenance of the stemness. Whether the germ lineage is specified early during embryogenesis or after gastrulation is still unknown. Similarly, Reitzel et al. (2016) found *nanos*, *piwi*, and *vasa* expressed during embryogenesis up to the cydippid stage, although with no overlapping expression in regions that contain germ cells or gametes. They did find expression of these germline genes in a region of high proliferation in the apical organ and tentacle apparatus of the cydippid larval stage, suggesting that this set of germline genes might be exclusively involved in stem cell specification and maintenance and that the germline is later specified or it employs a different molecular machinery.

4.7 ACOELA

Even though acoels were traditionally considered to be platyhelminths, the advent of molecular phylogenies quickly placed them as early bilaterians (Ruiz-Trillo et al., 1999). Despite later controversies (Hejnol and Pang, 2016; Philippe et al., 2011; Telford and Copley, 2016), the current understanding is that acoels, together with nemertodermatids and *Xenoturbella*—the Xenacoelomorpha—are the sister group of all other bilaterians—the Nephrozoa (Figure 4.1). From these three groups, acoels are the best studied, as well as the most diverse and broadly distributed. Acoels are free living, mostly marine worms. Many live in the benthic zone and typically feed on algae, detritus, or small particles. They lack a conventional gut and digest food by direct phagocytosis in a syncytial digestive system. Species such as *Symsagittifera roscoffensis* establish symbiotic relationships with photosynthetic algae (Bailly et al., 2014).

After the proposal of the Xenacoelomorpha as the most early-branching bilaterians, their main evolutionary novelty is the presence of a mesoderm (Hejnol and Pang, 2016). This tissue, also known as parenchyma or mesenchyme, gives rise to the musculature (Chiodin et al., 2011, 2013) but also contains a population of undifferentiated stem cells, called neoblasts, and the gonads. All of these cell types share the expression of mesodermal genes (Chiodin et al., 2013). Acoels are capable of whole-body regeneration that is controlled through ancient developmental pathways such as Wnt for anterior-posterior axis regeneration and BMP-Admp for dorsal-ventral axis regeneration (Srivastava et al., 2014).

Most acoels reproduce sexually. They are hermaphroditic—though some species can alternate sexes—and typically cross-fertilize. The gonads in acoels are not separated from the rest of the parenchymal tissue by any structure. Their position in the animal is highly variable, and often the gonads are unpaired, showing an asymmetric distribution. However, some species exploit their high regenerative capacities to reproduce asexually. For instance, species of the *Convolutriloba* and *Paratomella* genera reproduce by budding and paratomy (Achatz et al., 2013; Bourlat and Hejnol, 2009; Sikes and Bely, 2010). Regeneration in acoels is driven by the neoblasts population. Acoel neoblasts show remarkable similarities with planarian neoblasts (reviewed by (Gehrke and Srivastava, 2016). Regeneration in the acoel species *Hofstenia miamia* is also similar to planarian regeneration in terms of signalling pathways (Srivastava et al., 2014) and patterning (Raz et al., 2017). Acoel stem cells are distributed throughout the mesenchyme of the animal, and, together with germ line cells, they are the only proliferating cell type. It is still unknown if acoel neoblasts bear nuage granule particles, although the gametes clearly show them (e.g., Falleni and Gremigni, 1990). However, both neoblasts and germ cells express GMP components. Expression of *piwi* has been studied in both *Isodiametra pulchra* (De Mulder et al., 2009) and *Hofstenia miamia* (Srivastava et al., 2014). De Mulder et al. (2009) pioneered the acoel neoblasts field at the molecular level showing that, similar to platyhelminth neoblasts, acoel neoblasts incorporate BrdU and are eliminated by hydroxyurea treatments as well as irradiation. Apart from these features, the molecular identity of acoel neoblasts is vastly unknown. The development of *H. miamia* as a new model species as well as the tools already present in *I. pulchra* foretell future studies on the nature of the acoel stem cell system.

The relationship between neoblasts and the germ line in acoels is still understudied. Both are present in the mesenchyme and express similar genes. De Mulder and colleagues (2009) showed that freshly hatched *I. pulchra* juveniles have putative primordial germ cells, again similar to the situation observed in planarians. This indicates that the germ cells are already specified during embryonic development. Nevertheless, in planarians, neoblasts always retain the capacity of regenerating the germ cells (Wang et al., 2007). It would be interesting to see if, similar to planarians, acoel neoblasts can also replace the germ cells. Further experiments are needed to understand the lineage relationship between the germ cells and the adult neoblasts.

4.8 PLATYHELMINTHES

Planarians regenerate well, being able to reassemble entire organisms from tissue fragments a mere 1/279th of their body size (Alvarado, 2000), and both asexual and sexual reproduction are used to maintain natural populations. A specific cell type, the neoblast, is supposed to be directly involved in such regenerative processes (Nieuwkoop and Sutasurya, 1981). Neoblasts are free parenchymal cells that are amoeboid, basophilic, nucleolated, and possess a scanty rim of cytoplasm filled with ribosomes and very few endoplasmic reticulum (Rinkevich et al., 2009). Neoblasts are considered to be undifferentiated pluripotent cells (Baguñá, 1981; Nieuwkoop and Sutasurya, 1981; Wagner et al., 2011), originated during embryogenesis, although they can also be produced *de novo* in adults (Woodruff and Burnett, 1965). Nuage-like granules among mitochondrial clusters (Figure 4.1B) can be found in their cytoplasm (Hay and Coward, 1975; Auladell et al., 1993; Isaeva et al., 2005), although such granules have also been described in other somatic cells of planarians (Sauzin, 1968). Interestingly, neoblasts are also characterized by alkaline phosphate activity (Lender and Gabriel, 1960) and are labeled with bromodeoxyuridine (Newmark and Sánchez Alvarado, 2000). Neoblasts have been long postulated as the origin of germ cells, at least in turbellarians (Nieuwkoop and Sutasurya, 1981), based on histological evidences. Importantly, amputated pieces of planarians that do not contain germ cells can regenerate the germline (Wang et al., 2007). Therefore, neoblasts may retain the capacity to regenerate the germline throughout the life cycle. Germline molecular markers have been isolated from planarians and their expression observed in neoblasts and germ cells: *vasa*-like genes (Shibata et al., 1999; Ohashi et al., 2007; Pfister et al., 2008) and *piwi* (Reddien et al., 2005; Rossi et al., 2006). Sato and coworkers (2006) isolated a *nanos*-related gene (*Djnos*) from *Dugesia japonica*, and showed its expression in both germ cells (spermatogonia and oogonia) and in a subpopulation of neoblasts found in asexual planarians that gave rise to germ cells during its sexualization process. They also demonstrated that neoblasts express *Djnos* upon hatching and also during regeneration of any piece of the adult body. In *Schmidtea mediterranea*, *nanos*-like gene *Smednos* is also expressed in germline and eye precursor cells during development and regeneration (Handberg-Thorsager and Saló, 2007) (Figure 4.3). The conclusion of Sato et al. (2006) implied that the mechanism of segregation in planarians might be somewhat similar to that of the mouse (i.e., induction), where early embryonic stem cells can give rise to all types of cells, including germline cells (see review in Sato et al., 2006). It is possible that this population also gives rise to the adult planarian neoblasts population, which retains germline potential (Wang et al., 2007).

Planarian germ cell development has been more recently studied in depth by the Newmark lab among others. These studies have shown that apart from many RNA-binding proteins shared with neoblasts, germ cell enriched transcripts are, for the major part, RNA-binding proteins (Wang et al., 2010) as well, but also include a transcription factor (NF-YB) that regulates spermatogonial stem cell self-renewal (Iyer et al., 2016). Germ cell development is controlled by peptide hormones, including several neuropeptides (Collins III et al., 2010) that likely induce their development according to environmental or physiological conditions.

4.9 CONCLUDING REMARKS

Regeneration is known to occur by two mechanisms: with and without cell proliferation. While sponges and cnidarians can regenerate without active cell proliferation, planarians and acoels usually do it through active cell proliferation (Alvarado, 2000). It is the current paradigm that the organism's ability to regenerate will lie on its ability to regulate cellular pluripotency (see Alvarado, 2000). Therefore, maintaining large populations of pluripotent stem cells, as sponges and cnidarians do, or the ability to deploy the necessary physiological mechanisms to develop a blastema, like acoels and planarians, will be behind their astonishing regeneration capacities. In this sense, the involvement of the conserved GMP in both germline segregation and maintenance of pluripotency is therefore an essential pillar for regeneration. However, the interplay of germline determination and pluripotency regulation in lineages with high regeneration capacities is far from being understood. Further investigation of the presence of nuage in the ctenophore and sponge cells, especially in choanocytes and sperm in the latter is needed to physically locate gamete precursors. Such an organelle is a common feature of all precursors of gametes across the metazoan lineages (Eddy, 1975), and it has been proved to contain many of the germline markers in other metazoans (Cinalli et al., 2008). Also, and more importantly, isolation of the germline genetic machinery and the unraveling of the expression patterns of germline genes during the embryogenesis and gametogenesis of a diverse array of ctenophore and sponge lineages is a key step for understanding the germline evolution of animals.

In 2003, Extavour and Akam considered that neither the cells nor the segregation mode could be considered homologous within Metazoa since many different cell types coming from different germ layers are involved in such processes in the different animal groups. But still, early-splitting lineages of metazoans are the obscure spot in the animal tree of life regarding these questions, and it is only the investigation of such early animal lineages that could shed light on both the acquisition of the germ lineage as a defined and early segregated cell line and its evolution. The presence of genes such as the GMP genes that are shared between germ line cells and pluripotent stem cells highlights the possible relationship that these two might have. It is still unknown if these represent two branches of the same cell lineage, different parts of a germline cycle, like the Primordial Stem Cell hypothesis suggests, or are completely independent from each other.

New technologies such as single-cell transcriptomics (Griffiths et al., 2018) will be key to answer these questions. These techniques allow scientists to obtain mRNA profiles of tens of thousands of cells and classify them in cell types. This has enabled the detailed molecular profiling of cell types in sponges, ctenophores, cnidarians, placozoans, and planarians (Sebé-Pedrós et al., 2018a,b; Plass et al., 2018; Fincher et al., 2018). Furthermore, differentiation trajectories can be reconstructed, in both embryos and adult organisms (Wagner et al., 2018; Farrell et al., 2018; Briggs et al., 2018; Plass et al., 2018). Combining these methods with CRISPR-Cas approaches allows us to directly trace lineages (reviewed in Kester and van Oudenaarden, 2018). Therefore, single cell transcriptomics methods open new promising avenues to elucidate the origin of germline cells in many animals and investigate its relationship with pluripotent stem cell populations.

ACKNOWLEDGEMENTS

We are indebted to Dr. Sally Leys and Dr. Andreas Hejnol for the invitation to contribute this review for the volume Evolution of Metazoan Cell Types and two anonymous reviewers whose insightful comments and helpful corrections have greatly improved this manuscript. This chapter is a contribution under the grant from the

European Union's Horizon 2020 research and innovation program under Grant Agreement No. 679849 ("SponGES"), a grant of the Spanish Ministry of Science and Innovation (PID2019-105769GB-I00), and the grant number 9278 from the Villum Foundation to AR.

REFERENCES

Achatz, J. G., Chiodin, M., Salvenmoser, W., Tyler, S., & Martinez, P. 2013. The Acoela: On their kind and kinships, especially with nemertodermatids and xenoturbellids (Bilateria incertae sedis). *Org. Divers. Evol.* 13: 267–286.

Akhmadieva, A. V. 2008. *Morphofunctional Study of Stem Cells in Invertebrates with Reproductive Strategy Including a Sexual Reproduction.* Extended Abstract of Cand. Sci. (Biol.) Dissertation.

Alié, A., Hayashi, T., Sugimura, I., Manuel, M., Sugano, W., Mano, A., Satoh, N., Agata, K., & Funayama, N. 2015. The ancestral gene repertoire of animal stem cells. *PNAS* 112(51): E7093–E7100.

Alié, A., Leclère, L., Jager, M., Dayraud, C., Chang, P., Le Guyader, H., Quéinnec, E., & Manuel, M. 2011. Somatic stem cells express *Piwi* and *Vasa* genes in an adult ctenophore: Ancient association of "germline genes" with stemness. *Dev. Biol.* 350(1): 183–197.

Alvarado, A. S. 2000. Regeneration in the metazoans: Why does it happen?. *Bioessays* 22(6): 578–590.

Auladell, C., García-Valero, J., & Baguñá, J. 1993. Ultrastructural localization of RNA in the chromatoid bodies of undifferentiated cells (neoblasts) in planarians by the RNase-Gold complex technique. *J. Morphol.* 216: 319–326.

Bailly, X., Laguerre, L., Correc, G., Dupont, S., Kurth, T., Pfannkuchen, A., Entzeroth, R., Probert, I., Vinogradov, S., Levhauve, C., Garet-Delmas, M. J., Reichert, & Hartenstein, V. 2014. The chimerical and multifaceted marine acoel *Symsagittifera roscoffensis*: from photosymbiosis to brain regeneration. *Front. Microbiol.* 5: 498.

Baguñá, J. 1981. Planarian neoblasts. *Nature* 290: 14–15.

Borojevic, R. 1969. Étude du développement et de la différentiation cellulaire d'éponges calcaires Calcinées (genres *Clathrina* et *Ascandra*). *Ann. Embryol. Morphog.* 2: 15–36.

Bosch, T. C. G., & David, C. N. 1987. Stem cells of *Hydra magnipapillata* can differentiate into somatic cells and germ line cells. *Dev. Biol.* 121: 182–191.

Bourlat, S. J., & Hejnol, A. 2009. Acoels. *Curr. Biol.* 19: R279–R280.

Bridge, D., Cunningham, C. W., DeSalle, R., & Buss, L. W. 1995. Class-level relationships in the phylum Cnidaria: molecular and morphological evidence. *Mol. Biol. Evol.* 12: 679–689.

Brien, P. and Reniers-Decoen, M. 1955. La signification de les cellules interstitielles de Hydres d'eau douce et le problem la réserve embryonaire. *Bull. Biol. Fr. Belg.* 89: 258–325.

Briggs, J. A., Weinreb, C., Wagner, D. E., Megason, S., Peshkin, L., Kirschner, M. W., & Klein, A. M. 2018. The dynamics of gene expression in vertebrate embryogenesis at single-cell resolution. *Science* 360(6392): eaar5780.

Brown, F. D., & Swalla, B. J. 2007. *Vasa* expression in a colonial ascidian, *Botrylloides violaceus*. *Evol. Dev.* 9: 165–177.

Brown, F. D., Tiozzo, S., Roux, M. M., Ishizuka, K., Swalla, B. J., & De Tomasso, A. W. 2009. Early lineage specification of long-lived germline precursors in the colonial ascidian *Botryllus schlosseri*. *Development* 136: 3485–3494.

Brown, S. J., Cole, M. D., & Erives, A. J. 2008. Evolution of the holozoan ribosome biogenesis regulon. *BMC Genomics* 9: 442.

Buss, L. W. 1983. Evolution, development and the units of selection. *PNAS* 80: 1387–1391.

Buss, L. W. 1988. Diversification and germ-line determination. *Paleobiology* 14: 313–321.

Carmell, M. A., Xuan, Z., Zhang, M. Q., & Hannon, G. J. 2002. The Argonaute family: Tentacles that reach into RNAi, developmental control, stem cell maintenance, and tumorigenesis. *Genes Dev.* 16: 2733–2742.

Carré, C., Djediat, C., & Sardet, C. 2002. Formation of a large Vasa-positive germ granule and its inheritance by germ cells in the enigmatic chaetognaths. *Development* 129: 661–670.

Cerutti, H., Mian, N., & Bateman, A. 2000. Domains in gene silencing and cell differentiation: The novel PAZ domain and redefinition of the Piwi domain. *Trends Biochem. Sci.* 25: 481–482.

Cinalli, R. M., Rangan, P., & Lehmann, R. 2008. Germ cells are forever. *Cell* 132: 559–562.

Chiodin, M., Børve, A., Berezikov, E., Ladurner, P., Martinez, P., & Hejnol, A. 2013. Mesodermal gene expression in the acoel *Isodiametra pulchra* indicates a low number of mesodermal cell types and the endomesodermal origin of the gonads. *PLoS One* 8(2): e55499.

Chiodin, M., Achatz, J. G., Wanninger, A., & Martinez, P. 2011. Molecular architecture of muscles in an acoel and its evolutionary implications. *J. Exp. Zool. B Mol. Dev. Evol.* 316: 427–439.

Collins III, J. J., Hou, X., Romanova, E. V., Lambrus, B. G., Miller, C. M., Saberi, A., Sweedler, J. V., & Newmark, P. A. 2010. Genome-wide analyses reveal a role for peptide hormones in planarian germline development. *PLoS Biol.* 8(10): e1000509.

Connes, R., Paris, J., & Artiges, J. M. 1974. L'origine des cellules blastogenetiques chez *Suberites domuncula* Nardo. L'équilibre choanocytes-archeocytes chez les spontiaires. *Ann. Sci. Natur. Zool. (Paris)* 16: 111–118.

Cox, D. N., Chao, A., & Lin, H. 2000. *piwi* encodes a nucleoplasmic factor whose activitity modulates the number and division rate of germline stem cells. *Development* 127: 503–514.

Degnan, B. M., Adamska, M., Richards, G. S., Larroux, C., Leininger, S., Bergum, B., Calcino, A., Taylor, K., Nakanishi, N., & Degnan, S. M. 2015. Porifera. In *Evolutionary Developmental Biology of Invertebrates 1*. Vienna: Springer, pp. 65–106.

De Mulder, K., Kuales, G., Pfister, D., Willems, M., Egger, B., Salvenmoser, W., Thaler, M., Gorny, A.-K., Hrouda, M., Borgonie, G., & Ladurner, P. 2009. Characterization of the stem cell system of the acoel Isodiametra pulchra. *BMC Dev. Biol.* 9: 69.

Dunn, C. W., Hejnol, A., Matus, D. Q., Pang, K., Browne, W. E., Smith, S. A., Seaver, E., Rouse, G. W., Obst, M., Edgecombe, G. D., Sorensen, M. V., Haddock, S. H. D., Schmidt-Rhaesa, A., Okusu, A., Kristensen, R. M., Wheeler, W. C., Martindale, M. Q., & Giribet, G. 2008. Broad phylogenomic sampling improves resolution of the animal tree of life. *Nature* 452: 745–749.

Eckelbarger, K. J., & Larson, R. 1992. Ultrastructure of the ovary and oogenesis in the jellyfish *Linuche unguiculata* and *Stomolophus meleagris*, with a review of ovarian structure in the Scyphozoa. *Mar. Biol.* 114: 633–643.

Eddy, E. M. 1975. Germ plasm and the differentiation of the germ cell line. *Int. Rev. Cytol.* 43: 229–280.

Eerkes-Medrano, D., Feehan, C. J., & Leys, S. P. 2015. Sponge cell aggregation: Checkpoints in development indicate a high level of organismal complexity. *Invertebr. Biol.* 134(1): 1–18.

Eitel, M., Guidi, L., Hadrys, H., Balsamo, M., & Schierwater, B. 2011. New insights into placozoan sexual reproduction and development. *PLoS One* 6(5): e19639.

Ephrussi, A., & Lehmann, R. 1992. Induction of germ cell formation by *oskar*. *Nature* 358: 387–392.

Ereskovsky, A. V. 2010. *The Comparative Embryology of Sponges.* Berlin: Springer Science & Business Media.

Ewen-Campen, B., Schwager, E. E., & Extavour, C. G. M. 2010. The molecular machinery of germ line specification. *Mol. Reprod. Dev.* 77: 3–18.

Extavour, C. G., Pang, K., Matus, D. Q., & Martindale, M. Q. 2005. *vasa* and *nanos* expression patterns in a sea anemone and the evolution of bilaterian germ cell specification mechanisms. *Evol. Dev.* 7: 201–215.

Extavour, C. G. M., & Akam, M. 2003. Mechanisms of germ cell specification across the metazoans: Epigenesis and preformation. *Development* 130: 5869–5884.

Falleni, A., & Gremigni, V. 1990. Ultrastructural study of oogenesis in the acoel turbellarian *Convoluta. Tissue Cell* 22(3): 301–310.

Farrell, J. A., Wang, Y., Riesenfeld, S. J., Shekhar, K., Regev, A., & Schier, A. F. 2018. Single-cell reconstruction of developmental trajectories during zebrafish embryogenesis. *Science* 360(6392): eaar3131.

Fautin, D. G., & Mariscal, R. N. 1991. Cnidaria: Anthozoa. In F. W. Harrison & J. A. Westfall (eds.), *Microscopic Anatomy of Invertebrates, Vol. 2: Placozoa, Porifera, Cnidaria, and Ctenophora.* New York, NY: Wiley-Liss, pp. 267–358.

Fiedler, K. A. 1888. Über Ei- und Samenbildung bei *Spongilla fluviatilis. Zetischr. Wiss. Zool.* 47: 85–128.

Fierro-Constaín, L., Schenkelaars, Q., Gazave, E., Haguenauer, A., Rocher, C., Ereskovsky, A., Borchielllini, C., & Renard, E. 2017. The conservation of the germline multipotency program, from sponges to vertebrates: A stepping stone to understanding the somatic and germline origins. *Genome Biol. Evol.* 9(3): 474–488.

Fincher, C. T., Wurtzel, O., de Hoog, T., Kravarik, K. M., & Reddien, P. W. 2018. Cell type transcriptome atlas for the planarian *Schmidtea mediterranea. Science* 360(6391): eaaq1736.

Finnerty, J. R., & Martindale, M. Q. 1999. Ancient origins of axial patterning genes: Hox and ParaHox genes in the Cnidaria. *Evol. Dev.* 1: 16–23.

Frank, U., Plickert, G., & Müller, W. A. 2009. Cnidarian interstitial cells: The dawn of stem cell research. In B. Rinkevich and V. Matranga (eds.), *Stem Cells in Marine Organisms.* Dordrecht: Springer.

Funayama, N. 2008. Stem cell system of sponge. In T. C. G. Bosch (ed.), *Stem Cells: From Hydra to Man.* Berlin: Springer Science and Business Media B.V., pp. 17–35.

Funayama, N. 2010. The stem cell system in demosponges: Insights into the origin of somatic stem cells. *Dev. Growth Differ.* 52: 1–14.

Funayama, N., Nakatsukasa, M., Mohri, K., & Agata, K. 2010. *Piwi* expression in archeocytes and choanocytes in demosponges: insights into the stem cell system in demosponges. *Evol. Dev.* 12: 275–287.

Funayama, N. 2018. The cellular and molecular bases of the sponge stem cell systems underlying reproduction, homeostasis and regeneration. *Int. J. Dev. Biol.* 62(6–7–8): 513–525.

Gaino, E., Burlando, B., Buffa, P., & Sarà, M. 1986. Ultrastructural study of spermatogenesis in *Oscarella lobularis* (Porifera, Demospongiae). *Int. J. Invertebr. Reprod. Dev.* 10: 297–305.

Galliot, B., & Schmid, V. 2003. Cnidarians as a model system for understanding evolution and regeneration. *Int. J. Dev. Biol.* 46(1): 39–48.

Gehrke, A. R., & Srivastava, M. 2016. Neoblasts and the evolution of whole-body regeneration. *Curr. Opin. Genet. Dev.* 40: 131–137.

Gilbert, S. F. 2000. *Developmental Biology.* Sunderland, MA: Sinauer Associates, Inc.

Girard, A., Sachidanandam, R., Hannon, G. J., & Carmell, M. A. 2006. A germline-specific class of small RNAs binds mammalian Piwi proteins. *Nature* 442: 199–202.

Gonobobleva, E. L., & Efremova, S. M. 2017. Germ cell determinants in the oocytes of freshwater sponges. *Russ. J. Dev. Biol.* 48(3): 231–235.

Green, K. J., & Kirk, D. L. 1981. Cleavage patterns, cell lineages, and development of a cytoplasmic bridge system in *Volvox* embryos. *J. Cell Biol.* 91: 743–755.

Grell, K. G., & Benwitz, G. 1981. Ergänzende Untersuchungen zur Ultrastruktur von *Trichoplax adhaerens* F. E. Schulze (Placozoa). *Zoomorphology* 98: 47–67.

Grell, K. G., & Ruthmann, A. 1991. Placozoa. In F. W. Harrison and J. A. Westfall (eds.), *Microscopic Anatomy of Invertebrates*. New York, NY: Wiley-Liss, pp. 13–27.

Griffiths, J. A., Scialdone, A., & Marioni, J. C. 2018. Using single-cell genomics to understand developmental processes and cell fate decisions. *Mol. Syst. Biol.* 14(4): e8046.

Grimson, A., Srivastava, M., Fahey, B., Woodcroft, B. J., Rosaria Chiang, H., King, N., Degnan, B. M., Rokhsar, D. S., & Bartel, D. P. 2008. Early origins and evolution of microRNAs and Piwi-interacting RNAs in animals. *Nature* 455: 1193–1198.

Gustafson, E. A., & Wessel, G. M. 2010. Vasa genes: Emerging roles in the germ line and in multipotent cells. *Bioessays* 32: 626–637.

Haeckel, E. 1872. *Die Kalkschwämme, eine Monographie*. Berlin: G. Reimer.

Handberg-Thorsager, M., & Saló, E. 2007. The planarian *nanos*-like gene *Smednos* is expressed in germline and eye precursor cells during development and regeneration. *Dev. Genes Evol.* 217: 403–411.

Hay, E. D., & Coward, S. J. 1975. Fine structure studies on the planarian, *Dugesia*: I. Nature of the "neoblast" and other cell types in noninjured worms. *J. Ultrastr. Res.* 50: 1–21.

Hayashi, Y., Hayashi, M., & Kobayashi, S. 2004. *Nanos* supresses somatic cell fate in *Drosophila* germ line. *PNAS* 101: 10338–10342.

Hegner, R. W. 1914. *The Germ-Cell Cycle in Animals*. New York, NY: The MacMillan Company.

Hejnol, A., & Pang, K. 2016. Xenacoelomorpha's significance for understanding bilaterian evolution. *Curr. Opin. Genet. Dev.* 39: 48–54.

Hejnol, A., Obst, M., Stamatakis, A., Ott, M., Rouse, G. W., Edgecombe, G. D., Martinez, P., Baguñà, J., Bailly, X., Jondelius, U., Wiens, M., Müller, W. E. G., Seaver, E., Wheeler, W. C., Martindale, M. Q., Giribet, G., & Dunn, C. W. 2009. Assessing the root of bilaterian animals with scalable phylogenomic methods. *Proc. R. Soc. Biol. Sci. Ser. B* 276: 4261–4270.

Holstein, T. W., Hobmayer, E., & Technau, U. 2003. Cnidarians: An evolutionarily conserved model system for regeneration? *Dev. Dyn.* 226(2): 257–267.

Ikenishi, K. 1998. Germ plasm in *Caenorhabditis elegans*, *Drosophila* and *Xenopus*. *Dev. Growth Differ.* 40: 1–10.

Isaeva, V. V., Akhmadieva, A. V., Aleksandrova, Y. N., & Shukalyuk, A. I. 2009. Morphofunctional organization of reserve stem cells providing for asexual and sexual reproduction of invertebrates. *Russ. J. Dev. Biol.* 40: 57–68.

Isaeva, V. V., Alexandrova, Y. N., & Reunov, A. 2005. Interaction between chromatoid bodies and mitochondria in neoblasts and gonial cells of the asexual and spontaneously sexualized planarian, *Girardia (Dugesia) tigrina*. *Invertebr. Reprod. Dev.* 48: 119–128.

Iyer, H., Collins III, J. J., & Newmark, P. A. 2016. NF-YB regulates spermatogonial stem cell self-renewal and proliferation in the planarian *Schmidtea mediterranea*. *PLoS Genet.* 12(6): e1006109.

Johnstone, O., & Lasko, P. 2001. Translational regulation and RNA localization in *Drosophila* oocytes and embryos. *Annu. Rev. Genet.* 35: 365–406.

Juliano, C. E., Swartz, S. Z., & Wessel, G. M. 2010. A conserved germline multipotency program. *Development* 137: 4113–4126.

Juliano, C. E., & Wessel, G. M. 2010. Versatile germline genes. *Science* 329: 640.

Kerner, P., Degnan, S. M., Marchand, L., Degnan, B. M., & Vervoort, M. 2011. Evolution of RNA-binding proteins in animals: Insights from genome-wide analysis in the sponge *Amphimedon queenslandica*. *Mol. Biol. Evol.* 28(8): 2289–2303.
Kester, L., & van Oudenaarden, A. 2018. Single-cell transcriptomics meets lineage tracing. *Cell Stem Cell.* doi:10.1016/j.stem.2018.04.014.
King, F. J., Szakmary, A., Cox, D. N., & Lin, H. 2001. Yb modulates the divisions of both germline and somatic stem cells through *piwi*- and *hh*-mediated mechanisms in the *Drosophila* ovary. *Mol. Cell* 7: 497–508.
Kirk, M. L., Ransick, A., McRae, S. E., & Kirk, D. L. 1993. The relationships between cell size and cell fate in *Volvox carteri*. *J. Cell Biol.* 123: 191–208.
Kleinenberg, N. 1872. *Hydra – eine anatomische-entwicklungs-geshichtliche Unttersuschung.* Leipzig: W. Englemann.
Kranz, A. M., Tollenaere, A., Norris, B. J., Degnan, B. M., & Degnan, S. M. 2010. Indetifying the germline in an equally cleaving mollusc: *Vasa* and *Nanos* expression during embryonic and larval development of the vetigastropod *Haliotis asinina*. *J. Exp. Zool. B. Mol. Dev. Evol.* 314B: 1–13.
Krasko, A., Lorenz, B., Batel, R., Schröder, H. C., Müller, I. M., & Müller, W. E. 2000. Expression of silicatein and collagen genes in the marine sponge *Suberites domuncula* is controlled by silicate and myotrophin. *Eur. J. Biochem.* 267(15): 4878–4887.
Kuramochi-Miyagawa, S., Kimura, T., Ijiri, T. W., Isobe, T., Asada, N., Fujita, Y., Ikawa, M., Iwai, N., Okabe, M., Deng, W., Lin, H., Matsuda, Y., & Nakano, T. 2004. Mili, a mammalian member of a *piwi* family gene, is essential for spermatogenesis. *Development* 131: 839–849.
Lanna, E., & Klautau, M. 2010. Oogenesis and spermatogenesis in *Paraleucilla magna* (Porifera, Calcarea). *Zoomorphology* 129: 249–261.
Larkman, A. U. 1984. An ultrastructural study of the establishment of the testicular cysts during spermatogenesis in the sea anemone *Actinia fragacea* (Cnidaria: Anthozoa). *Gamete Res.* 9: 303–327.
Laumer, C. E., Gruber-Vodicka, H., Hadfield, M. G., Pearse, V. B., Riesgo, A., Marioni, J. C., & Giribet, G. 2018. Support for a clade of Placozoa and Cnidaria in genes with minimal compositional bias. *Elife* 7: e36278.
Lavrov, A. I., & Kosevich, I. A. 2016. Sponge cell reaggregation: Cellular structure and morphogenetic potencies of multicellular aggregates. *J. Exp. Zool. A Ecol. Genet. Physiol.* 325(2): 158–177.
Lehmann, R. 2016. Germ plasm biogenesis—An Oskar-centric perspective. In *Current Topics in Developmental Biology*, Vol. 116. Cambridge, MA: Academic Press, pp. 679–707.
Leininger, S., Adamski, M., Bergum, B., Guder, C., Liu, J., Laplante, M., Bråte, J., Hoffmann, F., Fortunato, S., Jordal, S., Rapp, H. T., & Adamska, M. 2014. Developmental gene expression provides clues to relationships between sponge and eumetazoan body plans. *Nat. Commun.* 5: 3905.
Lender, T., & Gabriel, A. 1960. Étude histochimique des néoblastes de *Dugesia lugubris* (Turberllarié, Triclade) avant et pendant la régénération. *Bull. Soc. Zool. France* 85: 100–110.
Lesh-Laurie, G. E., & Suchy, P. E. 1991. Cnidaria: Scyphozoa and Cubozoa. In F. W. Harrison and J. A. Westfall (eds.), *Microscopic Anatomy of Invertebrates*. New York, NY: Wiley-Liss, pp. 185–266.
Leys, S. P., Cheung, E., & Boury-Esnault, N. 2006. Embryogenesis in the glass sponge *Oopsacas minuta*: Formation of syncytia by fusion of blastomeres. *Integr. Comp. Biol.* 46: 104–117.
Littlefield, C. L. 1991. Cell lineages in Hydra: Isolation and characterization of an interstitial stem cell restricted to egg production in *Hydra oligactis*. *Dev. Biol.* 143(2): 378–388.

Littlefield, C. L. 1994. Cell-cell interactions and the control of sex determination in hydra. *Semin. Dev. Biol.* 5: 13–20.
Maas, O. 1893. Die Embryonal-Entwicklung und metamorphose der Cornacuspongien. *Zool. Jahrb.* 7: 331–448.
Maldonado, M., Bergquist, P. R., & Young, C. M. 2002. Phylum Porifera. In *Atlas of Marine Invertebrate Larvae*. New York, NY: Academic Press, pp. 21–50.
Maldonado, M. & Riesgo, A. 2008. Reproduction in Porifera: A synoptic overview. *Boletín de la Sociedad Catalana de Biología* 59: 29–49.
Martindale, M. Q., & Henry, J. Q. 1999. Intracellular fate mapping in a basal metazoan, the ctenophore *Mnemiopsis leidyi*, reveals the origins of mesoderm and the existence of indeterminate cell lineages. *Dev. Biol.* 214: 243–257.
Martindale, M. Q., Pang, K., & Finnerty, J. R. 2004. Investigating the origins of triploblasty: "mesodermal" gene expression in a diploblastic animal, the sea anemone *Nematostella vectensis* (phylum, Cnidaria; class, Anthozoa). *Development* 131: 2463–2474.
McLaren, A. 2003. Primordial germ cells in the mouse. *Dev. Biol.* 262: 1.15.
Megosh, H. B., Cox, D. N., Campbell, C., & Lin, H. 2006. The role of PIWI and the miRNA machinery in *Drosophila* germline determination. *Curr. Biol.* 16: 1884–1894.
Miller, S. M., & Kirk, D. L. 1999. *glsA*, a *Volvox* gene required for asymmetric division and germ cell specification, encodes a chaperone-like protein. *Development* 126: 649–658.
Minchin, E. A. 1900. Porifera. In E. R. Lankester (ed.), *A Treatise on Zoology*. Pt. 2. London: A. C. Black, pp. 1–178.
Mochizuki, K., Nishimiya-Fujisawa, C., & Fujisawa, T. 2001. Universal occurrence of the *vasa*-related genes among metazoans and their germline expression in *Hydra*. *Dev. Genes Evol.* 211: 299–308.
Mochizuki, K., Sano, H., Kobayashi, S., Nishimiya-Fujisawa, C., & Fujisawa, T. 2000. Expression and evolutionary conservation of *nanos*-related genes in *Hydra*. *Dev. Genes Evol.* 210: 591–602.
Moroz, L. L. 2015. Convergent evolution of neural systems in ctenophores. *J. Exp. Biol.* 218(4): 598–611.
Morrison, H. G., McArthur, A. G., Gillin, F. D., Aley, S. B., Adam, R. D., Olsen, G. J., Best, A. A., Cande, W. Z., Chen, F., Cipriano, M. J., Davids, B. J., Dawson, S. C., Elmendorf, H. G., Hehl, A. B., Holder, M. E., Huse, S. M., Kim, U. U., Lasek-Nesselquist, E., Manning, G., Nigam, A., Nixon, J. E. J., Palm, D., Passamaneck, N. E., Prabhu, A., Reich, C. I., Reiner, D. S., Samuelson, J., Svard, S. G., & Sogin, M. L. 2007. Genomic minimalism in the early diverging intestinal parasite *Giardia lamblia*. *Science* 317: 1921–1926.
Müller, W. E. G. 2006. The stem cell concept in sponges (Porifera): Metazoan traits. *Semin. Cell Dev. Biol.* 17: 481–491.
Newmark, P. A., & Sánchez Alvarado, A. 2000. Bromodeoxyuridine specifically labels the regenerative stem cells of planarians. *Dev. Biol.* 220: 142–153.
Niewkoop, P. D., & Sutasurya, L. A. 1981. *Primordial Germ Cells in the Invertebrates*. Cambridge: Cambridge University Press.
Nishimiya-Fujisawa, C., & Sugiyama, T. 1993. Genetic analysis of developmental mechanisms in *Hydra*: XX. Cloning of interstitial stem cells restricted to the sperm differentiation pathway in *Hydra magnipapillata*. *Dev. Biol.* 157: 1–9.
Nishimiya-Fujisawa, C., & Kobayashi, S. 2012. Germline stem cells and sex determination in *Hydra*. *Int. J. Dev. Biol.* 56(6-7-8): 499–508.
Noda, K., & Kanai, C. 1977. An ultrastructural observation of *Pelmatohydra robusta* at sexual and asexual stages, with special reference to "Germinal plasm". *J. Ultrastr. Res.* 61: 284–294.
Ohashi, H., Umeda, N., Hirazawa, N., Ozaki, Y., Miura, C., & Miura, T. 2007. Expression of vasa (vas)-related genes in germ cells and specific interference with gene functions

by double-stranded RNA in the monogenean, *Neobenedenia girellae*. *Int. J. Parasitol.* 37: 515–523.

Pang, K., & Martindale, M. Q. (2008). Comb jellies (Ctenophora): A model for basal metazoan evolution and development. *Cold Spring Harb. Protoc.* 2008(11): pdb-emo106.

Perović-Ottstadt, S., Cetkovic', H., Gamulin, V., Schröder, H. C., Kropf, K., Moss, C., Korzhev, M., Diehl-Seifert, B., Müller, I. M., & Müller, W. E. G. 2004. Molecular markers for germ cell differentiation in the demosponge *Suberites domuncula*. *Int. J. Dev. Biol.* 48: 293–305.

Peters, B., Casey, J., Aidley, J., Zohrab, S., Borg, M., Twell, D., & Brownfield, L. 2017. A conserved cis-regulatory module determines germline fate through activation of the transcription factor DUO1 promoter. *Plant Physiol.* 173(1): 280–293.

Pfister, D., De Mulder, K., Hartenstein, V., Kuales, G., Borgonie, G., Marx, F., Morris, J., & Ladurner, P. 2008. Flatworm stem cells and the germ line: Developmental and evolutionary implications of macvasa expression in *Macrostomum lignano*. *Dev. Biol.* 319: 146–159.

Philippe, H., Brinkmann, H., Copley, R. R., Moroz, L. L., Nakano, H., Poustka, A. J., Wallberg, A., Peterson, K. J., & Telford, M. J. 2011. Acoelomorph flatworms are deuterostomes related to *Xenoturbella*. *Nature* 470: 255–258.

Pick, K. S., Philippe, H., Schreiber, F., Erpenbeck, D., Jackson, D. J., Wrede, P., Wiens, M., Alié, A., Morgenstern, B., Manuel, M., & Wörheide, G. 2010. Improved phylogenomic taxon sampling noticeably affects nonbilaterian relationships. *Mol. Biol. Evol.* 27: 1983–1987.

Plass, M., Solana, J., Wolf, F. A., Ayoub, S., Misios, A., Glažar, P., Obermayer, B., Theis, F. J., Kocks, C., & Rajewsky, N. 2018. Cell type atlas and lineage tree of a whole complex animal by single-cell transcriptomics. *Science* 360(6391): eaaq1723.

Rabinowitz, J. S., Chan, X. Y., Kingsley, E. P., Duan, Y., & Lambert, J. D. 2008. Nanos is required in somatic blast cell lineages in the posterior of a mollusk embryo. *Curr. Biol.* 18: 331–336.

Raz, E. 2000. The function and regulation of *vasa*-like genes in germ-cell development. *Genome Biol.* 1: 1017.1–1017.6.

Raz, A. A., Srivastava, M., Salvamoser, R., & Reddien, P. W. 2017. Acoel regeneration mechanisms indicate an ancient role for muscle in regenerative patterning. *Nat. Commun.* 8: 1260.

Rebscher, N., Volk, C., Teo, R., & Plickert, G. 2008. The germ plasm component Vasa allows tracing of the interstitial stem cells in the cnidarian *Hydractinia echinata*. *Dev. Dyn.* 237: 1736–1745.

Rebscher, N., Zelada-González, F., Barnisch, T. U., & Arendt, D. 2007. Vasa unveils a common origin of germ cells and of somatic stem cells from the posterior growth zone in the polychaete *Platynereis dumerilii*. *Dev. Biol.* 306: 599–611.

Reddien, P. W., Oviedo, N. J., Jennings, J. R., Jenkin, J. C., & Alvarado, A. S. 2005. SMEDWI-2 is a PIWI-like protein that regulates planarian stem cells. *Science* 310: 1327–1330.

Reinke, V., Smith, H. E., Nance, J., Jones, S. J. M., Davis, E. B., Scherer, S., Ward, S., & Kim, S. K. 2000. A global profile of germline gene expression in *C. elegans*. *Mol. Cell* 6: 605–616.

Reitzel, A. M., Pang, K., & Martindale, M. Q. 2016. Developmental expression of "germline"- and "sex determination"-related genes in the ctenophore *Mnemiopsis leidyi*. *EvoDevo* 7(1): 17.

Riesgo, A., Farrar, N., Windsor, P. J., Giribet, G., & Leys, S. P. 2014. The analysis of eight transcriptomes from all poriferan classes reveals surprising genetic complexity in sponges. *Mol. Biol. Evol.* 31(5): 1102–1120.

Riesgo, A., & Maldonado, M. 2009. *Sexual Reproduction of Demosponges. Ecological and Evolutionary Implications*. Deutschland: Verlag Dr. Müller, Saarbrücken.

Riesgo, A., Maldonado, M., & Durfort, M. 2007b. Dynamics of gametogenesis, embryogenesis, and larval release in a Mediterranean homosclerophorid demosponge. *Mar. Freshw. Res.* 58: 398–417.

Riesgo, A., Taylor, C., & Leys, S. 2007a. Reproduction in a carnivorous sponge: The significance of the absence of an aquiferous system to the sponge body plan. *Evol. Dev.* 9: 618–631.

Rinkevich, Y., Matranga, V., & Rinkevich, B. 2009. Stem cells in aquatic invertebrates: Common premises and emerging unique themes. In B. Rinkevich and V. Matranga (eds.), *Stem Cells in Marine Organisms*. Dordrecht: Springer.

Ruiz-Trillo, I., Riutort, M., Littlewood, D. T. J., Herniou, E. A., & Baguñà, J. 1999. Acoel flatworms: Earliest extant bilaterian metazoans, not members of Platyhelminthes. *Science* 283: 1919–1923.

Rossi, L., Salvetti, A., Lena, A., Batistoni, R., Deri, P., Pugliesi, C., Loreti, E., & Gremigni, V. 2006. *DjPiwi-1*, a member of the *PAZ-Piwi* gene family, defines a subpopulation of planarian stem cells. *Dev. Genes Evol.* 216: 335–346.

Ryan, J. F., Pang, K., Mullikin, J. C., Martindale, M. Q., & Baxevanis, A. D. 2010. The homeodomain complement of the ctenophore *Mnemiopsis leidyi* suggests that Ctenophora and Porifera diverged prior to the ParaHoxozoa. *EvoDevo* 1(1): 9.

Ryan, J. F., Pang, K., Schnitzler, C. E., Nguyen, A. D., Moreland, R. T., Simmons, D. K., Koch, B. J., Francis, W. R., Havlak, P., NISC Comparative Sequencing Program, Smith, S. A., Putnam, N. H., Haddock, S. H. D., Dunn, C. W., Wolfsberg, T. G., Mullikin, J. C., Martindale, M. Q., & Baxevanis, A. D. 2013. The genome of the ctenophore *Mnemiopsis leidyi* and its implications for cell type evolution. *Science* 342(6164): 1242592.

Saffman, E. E., & Lasko, P. 1999. Germline in vertebrates and invertebrates. *Cell Mol. Life Sci.* 55: 1141–1163.

Sará, M. 1974. Sexuality in the Porifera. *Boll. Zool.* 41: 327–348.

Sato, K., Hayashi, Y., Ninomiya, Y., Shigenobu, S., Arita, K., Mukai, M., & Kobayashi, S. 2007. Maternal Nanos represses hid/skl-dependent apoptosis to maintain the germ line in *Drosophila* embryos. *PNAS* 104: 7455–7460.

Sato, K., Shiata, N., Orii, H., Amikura, R., Sakurai, T., Agata, K., Kobayashi, S., & Watanabe, K. 2006. Identification and origin of the germline stem cells as revelealed by the expression of *nanos*-related gene in planarians. *Dev. Growth Differ.* 48: 615–628.

Sauzin, M. J. 1968. Présence d'emissions nucléaires dans les cellules différenciées et en différenciation de la planaire adulte *Dugesia gonocephala*. *C. R. Acad. Sci. Paris* 267: 1146–1148.

Schierwater, B., de Jong, D., & DeSalle, R. 2009b. Placozoa and the evolution of Metazoa and intrasomatic differentiation. *Int. J. Biochem. Cell Biol.* 41: 370–379.

Schierwater, B., Eitel, M., Jakob, W., Osigus, H.-J., Hadrys, H., Dellaporta, S. L., Kolokotronis, S.-O., & DeSalle, R. 2009a. Concatenated analysis sheds light on early metazoan evolution and fuels a modern "Urmetazoon" hypothesis. *PLoS Biol.* 7: e1000020.

Schulze, F. E. 1871. *Uber den Bau und die Entzvicklung von Cordylophora lacustris (Allman)*. Leipzig: Leipzig Publisher.

Schulze, F. E. 1875. Ueber den Bau und die Entwicklung von *Sycandra raphanus* Haeckel. *Zetischr. Wiss. Zool.* 4: 31262–31295.

Schüpbach, T., & Wieschaus, E. 1986. Maternal-effect mutations altering the anterior-posterior pattern of the *Drosophila* embryo. *Dev. Genes Evol.* 195: 302–317.

Sebé-Pedrós, A., Chomsky, E., Pang, K., Lara-Astiaso, D., Gaiti, F., Mukamel, Z., Amit, I., Hejnol, A., Degnan, B. M., & Tanay, A. 2018a. Early metazoan cell type diversity and the evolution of multicellular gene regulation. *Nat. Ecol. Evol.* 1: 1176–1188.

Sebé-Pedrós, A., Saudemont, B., Chomsky, E., Plessier, F., Mailhé, M. P., Renno, J., Loe-Mie, Y., Lifshitz, A., Mukamel, Z., Schmutz, S., Novault, S., Steinmetz, P. R. H., Spitz, F.,

Tanay, A., & Marlow, H. 2018b. Cnidarian cell type diversity and regulation revealed by whole-organism single-cell RNA-Seq. *Cell* 173(6): 1520–1534.

Seipel, K., Yanze, N., & Schmid, V. 2004. The germ line and somatic stem cell gene Cniwi in the jellyfish *Podocoryne carnea*. *Int. J. Dev. Biol.* 48: 1–7.

Seydoux, G., & Braun, R. E. 2006. Pathway to totipotency: Lessons from germ cells. *Cell* 127: 891–904.

Shibata, N., Umesono, Y., Orii, H., & Sakurai, T. 1999. Expression of *vasa(vas)*-related genes in germline cells and totipotent somatic stem cells of planarians. *Dev. Biol.* 206: 73–87.

Sidall, M. E., Martin, D. S., Bridge, D., Desser, S. S., & Cone, D. K. 1995. The demise of a phylum of protists: Phylogeny of Myxozoa and other parasitic Cnidaria. *J. Parasitol.* 81: 961–967.

Siebert, S., Goetz, F. E., Church, S. H., Bhattacharyya, P., Zapata, F., Haddock, S. H., & Dunn, C. W. 2015. Stem cells in *Nanomia bijuga* (Siphonophora), a colonial animal with localized growth zones. *EvoDevo* 6(1): 22.

Siebert, S., & Juliano, C. E. 2017. Sex, polyps, and medusae: Determination and maintenance of sex in cnidarians. *Mol. Reprod. Dev.* 84(2): 105–119.

Sikes, J. M., & Bely, A. E. 2010. Making heads from tails: Development of a reversed anterior-posterior axis during budding in an acoel. *Dev. Biol.* 338: 86–97.

Signorovitch, A. Y., Dellaporta, S. L., & Buss, L. W. 2005. Molecular signatures for sex in the Placozoa. *PNAS* 102: 15518–15522.

Simpson, T. L. 1984. Gamete, embryo, larval development. In *The Cell Biology of Sponges*. Berlin: Springer Verlag, pp. 341–413.

Solana, J. 2013. Closing the circle of germline and stem cells: The Primordial Stem Cell hypothesis. *Evodevo* 4(1): 2.

Srivastava, M., Mazza-Curll, K. L., van Wolfswinkel, J. C., & Reddien, P. W. 2014. Whole-body acoel regeneration is controlled by Wnt and Bmp-Admp signaling. *Curr. Biol.* 24: 1107–1113.

Srivastava, M., Begovic, E., Chapman, J., Putnam, N. H., Hellsten, U., Kawashima, T., Kuo, A., Mitros, T., Salamov, A., Carpenter, M. L., Signorovitch, A. Y., Moreno, M. A., Kamm, K., Grimwood, J., Schmutz, J., Shapiro, H., Grigoriev, I. V., Buss, L. W., Schierwater, B., Dellaporta, S. L., & Rokhsar, D. S. 2008. The *Trichoplax* genome and the nature of placozoans. *Nature* 454: 955–960.

Steeves, T. A., & Sussex, I. M. 1989. *Patterns in Plant Development*. Cambridge: Cambridge University Press.

Strome, S., & Updike, D. 2015. Specifying and protecting germ cell fate. *Nat. Rev. Mol. Cell Biol.* 16(7): 406.

Styhler, S., Nakamura, A., & Lasko, P. 2002. VASA localization requires the SPRY-domain and SOCS-box containing protein, GUSTAVUS. *Dev. Cell* 3: 865–876.

Sugiyama, T., & Sugimoto, N. 1985. Genetic analysis of developmental mechanics in hydra. XI. Mechanism of sex reversal by heterosexual parabiosis. *Dev. Biol.* 110: 413–421.

Swartz, S. Z., Chan, X. Y., & Lambert, J. D. 2008. Localization of Vasa mRNA during early cleavage of the snail *Ilyanassa*. *Dev. Genes Evol.* 218: 107–113.

Syed, T., & Schierwater, B. 2002a. *Trichoplax adhaerens*: Discovered as a missing link, forgotten as a hydrozoan, re-discovered as a key to metazoan evolution. *Viet Milieu* 52: 177–187.

Syed, T., & Schierwater, B. 2002b. The evolution of the Placozoa: A new morphological model. *Sencken. Lethaea* 82: 315–324.

Takeda, S., & Paszkowski, J. 2006. DNA methylation and epigenetic inheritance during plan gametogenesis. *Chromosoma* 115: 27–35.

Telford, M. J., & Copley, R. R. 2016. Zoology: War of the worms. *Curr. Biol.* 26: R335–R337.

Torras, R., Yanze, N., Schmid, V., & González-Crespo, S. 2004. *nanos* expression at the embryonic posterior pole and the medusa phase in the hydrozoan *Podocoryne carnea*. *Evol. Dev.* 6: 362–371.

Uriz, M. J., Turon, X., & Becerro, M. A. 2001. Morphology and ultrastructure of the swimming larvae of *Crambe crambe* (Demospongiae, Poecilosclerida). *Invertebr. Biol.* 120: 295–307.

Wagner, D. E., Weinreb, C., Collins, Z. M., Briggs, J. A., Megason, S. G., & Klein, A. M. 2018. Single-cell mapping of gene expression landscapes and lineage in the zebrafish embryo. *Science* 360(6392): 981–987.

Wagner, D. E., Wang, I. E., & Reddien, P. W. 2011. Clonogenic neoblasts are pluripotent adult stem cells that underlie planarian regeneration. *Science* 332(6031): 811–816.

Wang, Y., Stary, J. M., Wilhelm, J. E., & Newmark, P. A. 2010. A functional genomic screen in planarians identifies novel regulators of germ cell development. *Genes Dev.* 24(18): 2081–2092.

Wang, Y., Zayas, R. M., Guo, T., & Newmark, P. A. 2007. *nanos* function is essential for development and regeneration of planarian germ cells. *PNAS* 104: 5901–5906.

Weltner, W. 1907. Spongillidenstudien V. Zur biologie von *Ephydatia fluviatilis* und die bedeutung der amoebocyten für die Spongilliden. *Arch. Natur.* 73: 273–286.

Woodruff, L. S., & Burnett, A. L. 1965. The origin of the blastemal cells in *Dugesia tigrina*. *Exp. Cell Res.* 38: 295–305.

Wylie, C. 1999. Germ cells. *Cell* 96: 165–174.

Zapata, F., Goetz, F. E., Smith, S. A., Howison, M., Siebert, S., Church, S. H., Sanders, S. M., Ames, C. L., McFadden, C. S., France, S. C., Daly, M., Collins, A. G., Haddock, S. H. D., Dunn, C. W., & Cartwright, P. 2015. Phylogenomic analyses support traditional relationships within Cnidaria. *PLoS One* 10(10): e0139068.

5 Origin and Evolution of Epithelial Cell Types

Emmanuelle Renard, André Le Bivic, and Carole Borchiellini

CONTENTS

- 5.1 Introduction 76
  - 5.1.1 Epithelia Are Not a Prerogative of Animals 76
  - 5.1.2 The Definition of Epithelium in Metazoa 76
  - 5.1.3 The Classification of Animal Epithelial Types 78
  - 5.1.4 The Definition of Junction Types 78
  - 5.1.5 Cell Polarity Complexes and Cell Domains 79
  - 5.1.6 Do All Metazoans Have (the Same) Epithelia? 80
- 5.2 The Origin of Animal Epithelia 82
  - 5.2.1 Some Genes Encoding Extracellular Matrix Proteins Predate the Emergence of Animal Epithelia 83
  - 5.2.2 Integrins: An Ancient Protein Family 83
  - 5.2.3 Genes Encoding Cell-Cell Adhesive Proteins Predate the Emergence of Animal Epithelia 84
  - 5.2.4 Do Some Genes Encoding Members of Cell Polarity Complexes Predate the Emergence of Animal Epithelia? 85
  - 5.2.5 Main Conclusion: Cell Adhesion and Polarity Emerged Several Times during the Evolution of Holozoa 85
- 5.3 The Probably Ancestral Features of the Metazoan Epithelium 86
  - 5.3.1 The Composition of Ancestral Basement Membrane 86
  - 5.3.2 The Ancestral Toolkit to Achieve Cell Polarity Depends on Non-Bilaterian Relationships 87
  - 5.3.3 Were Adhesive Junctions Present Ancestrally? 89
    - 5.3.3.1 Communicating Junctions 89
    - 5.3.3.2 Sealing/Occluding Junctions 89
    - 5.3.3.3 Adhesive Junctions 91
  - 5.3.4 Epithelial Regulatory Signatures 91
- 5.4 Conclusions and Future Challenges 92
- Acknowledgements 93
- References 93

5.1 INTRODUCTION

5.1.1 Epithelia Are Not a Prerogative of Animals

Multicellularity emerged several times independently during the evolution of life (Abedin and King, 2010; Hinman and Cary, 2017; King, 2004; King et al., 2003; Niklas, 2014; Parfrey and Lahr, 2013), and so did epithelial-like tissues (Dickinson et al., 2011, 2012; Ganot et al., 2015; Maizel, 2018; Miller et al., 2013). In cell biology (Lowe and Anderson, 2015), epithelia are defined generally as a sheet of cells tightly bound together and capable of coordinated movements during morphogenesis. In the different lineages where they emerged, they usually shape and line organs (when present), cavities, and external borders; they control molecule and ion exchanges between the body and the environment and/or between different compartments of the body involved in vital physiological processes.

Because multicellularity and epithelial-like structures emerged several times independently, the molecules involved in their composition and patterning are different (Kania et al., 2014; Reynolds, 2011). Nevertheless, very ancient proteins predating multicellularity can play a key role in the establishment of epithelial features (Nagawa et al., 2010), such as Rho/ROP GTPase for polarity, because of their conserved involvement in controlling cytoskeleton and vesicular trafficking, and other ancient proteins can have been co-opted independently to perform a quite similar result, such as catenins in Amoebozoa and Metazoa (see section 2).

After this short synopsis, this chapter will focus on the origin and evolution of animal epithelia only.

5.1.2 The Definition of Epithelium in Metazoa

Epithelia of animals are usually considered as one of the four fundamental tissue types along with connective, nervous, and muscular tissues (Lowe and Anderson, 2015; Yathish and Grace, 2018). Epithelial tissues cover all the surfaces of the body exposed to the external environment and line organs and body cavities (named in this case: Covering and lining epithelia) hence providing protection and compartmentalization of the body. Epithelia also form much of the glandular tissue of an animal body (in this case, they are named glandular epithelia). During embryological development, epithelia are patterned early, after cleavage (Kim et al., 2017). Indeed, the organization of cells within tissues is the first sign of cell differentiation during embryogenesis, and the epithelial blastoderm is the starting point for morphogenesis during the development of a wide range of metazoans (Pozzi et al., 2017). It is only from epithelium that mesenchyme arises by epithelial-mesenchymal transition (EMT) during gastrulation (Kim et al., 2017). In bilaterian adults, epithelial cells derive from all three major embryonic layers (endo-, meso-, and ectoderm): for example, in vertebrates, the skin, part of the mouth and nose, and the anus develop from the ectoderm; cells lining the airways and most of the digestive system originate from the endoderm, while the epithelium lining vessels derive from the mesoderm. Epithelia are fundamental structures controlling permeability and allowing selective transfer of molecules between the animal and its environment and between body compartments.

Origin and Evolution of Epithelial Cell Types

Epithelial tissues also provide protection from physical, chemical, and biological agents and allow for coordinated tissue movements. Consequently, disruptions in epithelial properties cause developmental defects and are responsible for diseases in adult tissues (Miller et al., 2013; Royer and Lu, 2011; Sekiguchi and Yamada, 2018).

Currently, as for many other animal morpho-anatomical features, the histological characteristics defining an epithelium were postulated according to what was found in model animals, all pertaining to the bilaterian taxa, such as mammalian epithelial cells and *Drosophila* epithelial tissues (for review see Tyler, 2003). From these studies, epithelia have been defined as layers of cells showing coordinated polarity, with differences in structure and function between the apical side facing the external medium (or internal cavities) and a basal side in contact with a basement membrane (BM) made of collagen IV (Figure 5.1) (Rodriguez-Boulan and Nelson,

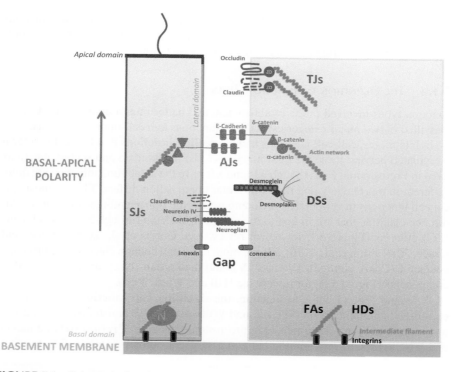

FIGURE 5.1 Schematic drawing summarizing the three features defining animal epithelia: 1) **Cell polarity** (basal-apical polarity: cell can harbor different features along this polarity axis, for instance, cilia in some cells); 2) **Cell-cell and cell-matrix junctions** found in Bilaterians and their core molecular composition (here, only structural components are noted rather than all the components needed for their trafficking and patterning). Left, junction types found in protostomes and non-chordates: Septate Junctions (SJs). Right, junction types found in vertebrates: Desmosomes (DSs), Tight Junctions (TJs). In between and on both sides are junction types found in all bilaterians: focal Adhesions (FAs), Hemi-Desmosomes (HDs), Gap junctions, Adhaerens Junctions (AJs); 3) **Basement membrane**, a dense sheet of extracellular matrix proteins providing support and increasing cohesiveness of cells.

1989). In addition, epithelial cells exhibit cell-cell junctions and cell-matrix contacts (Figure 5.1) which maintain cohesion between cells during morphogenesis processes and allow the coordination of cell movements by providing a seal between cells (Jefferson et al., 2004).

5.1.3 The Classification of Animal Epithelial Types

Despite this general definition, epithelia can have different organizations and structures. In addition to the distinction based on their general role (compartmentalizing *versus* glandular epithelia), epithelial tissues are identified by both the number of layers (simple, stratified, or pseudostratified) and the shape of the cells (squamous, cuboidal, or columnar) (summarized in Table 5.1, see Lowe and Anderson, 2015 for more details).

These different kinds of epithelial tissue generally perform different functions (absorption, regulation, excretion, filtration, secretion, protection, detection) according to their body position and their structural organization.

5.1.4 The Definition of Junction Types

Junction types involved in the establishment and patterning of epithelia are usually classified in two major categories: cell-cell junctions and cell-matrix junctions.

In bilaterians, cell-cell junctions include **sealing junctions** that link cells to form a regulated barrier: these are tight junctions (TJs) in Chordata and septate junctions (SJs) in Protostomia. Even though TJs and SJs have the same function, they are quite different in morphology and their molecular composition differs. TJs, the most apical cell junctions in vertebrates, are composed of the transmembrane proteins occludin and claudins that are linked to the actin cytoskeleton through zonula occludens (ZO) proteins (Figure 5.1, for recent reviews see Garcia et al., 2018; Yathish and Grace, 2018; Zihni et al., 2016). SJs are more basal and contain claudin-like proteins, contactin (Cont), neurexin IV (Nrx IV), and neuroglian (Nrg) among many other components (Figure 5.1) (for review see Hall and Ward, 2016).

A second type of cell-cell junction, the **signaling gap junction,** allows direct cell-to-cell communication (Skerrett and Williams, 2017; Yathish and Grace, 2018) and is composed in vertebrate and in prostostome lineages of respectively connexin and innexin proteins, two similar but non homologous proteins (Figure 5.1). In other

TABLE 5.1
General Classification of Animal Epithelia According to Their Organization

Number of cell layers		Cell shape			
		Flat	Cube	Parallel-piped	Irregular
	1	Simple squamous	Simple cuboidal	Simple columnar	Pseudostratified
	2 or more	Stratified squamous	Stratified cuboidal	Stratified columnar	

words, two distinct lineages found a convergent solution to solve the same problem of intercellular communication (Alexopoulos et al., 2004; Skerrett and Williams, 2017).

The third type of cell-cell junction, the **anchoring junction,** links cells and includes adhaerens junctions (AJs) and desmosomes (DSs). Unlike DSs, which contain the desmosomal transmembrane cadherin desmogleins and cytosolic desmoplakins (Johnson et al., 2014; Magie and Martindale, 2008; Yathish and Grace, 2018) that are restricted to vertebrates and which provide a link to intermediate filaments through desmoplakin, AJs are present in all bilaterian taxa and comprise classical cadherins which interact with the actin cytoskeleton through α-, β-, and δ-catenins (Figure 5.1) (Garcia et al., 2018; Miller et al., 2013; Yathish and Grace, 2018).

Finally, cell-matrix junctions enable the attachment of epithelial cells to the underlying basement membrane *via* focal adhesions (FAs) and hemi-desmosomes (HDs). As well as cell-cell junctions, cell-matrix junctions are needed to achieve collective migration and coordination of various epithelial morphogenetic processes that are important during development, tissue shaping, and wound healing. Both FAs and HDs are highly specialized structures with interactions between integrins, fibrillar proteins of the extracellular matrix, and the internal intermediate filament or actin network of the cell (Figure 5.1) (De Pascalis and Etienne-Manneville, 2017; Magie and Martindale, 2008; Miller et al., 2018; Walko et al., 2015). While FAs are found in non-epithelial cells types such as neurons or fibroblasts (Fischer et al., 2019), hemidesmosomes (HD) are specific to epithelia (Walko et al., 2015).

It is generally accepted that the different types of junctions and organization of different junctions in cells have allowed the development and differentiation of tissue types, and thus, this variety has played a key role in the evolution of animal body plans (Abedin and King, 2010; Magie and Martindale, 2008).

5.1.5 Cell Polarity Complexes and Cell Domains

The precise locations of the previously defined junctions is one of the numerous demonstrations of cell polarity. This polarity is also obvious by the localization and orientation of other cell features such as the nucleus, cilia or other cell extensions, the cytoskeleton, and vesicular trafficking. The cell is generally divided into three domains: the basal domain facing the basement membrane, the apical domain facing the external medium or internal cavity (lumen), and in-between these two domains, lies the lateral domain, the extent of which depends on the shape of cells (Table 5.1, Figures 5.1 and 5.2). This polarized structural and ultrastructural organization of an epithelial cell is patterned by at least three protein complexes named polarity complexes, each composed of three interacting proteins (Figure 5.2) (Assémat et al., 2008; Le Bivic, 2013): the apical PAR complex is composed of atypical Protein Kinase C (aPKC), the Partition defective 3 (PAR3), and Partition defective 6 (PAR6); the apical Crumbs complex is made of CRUMBS (CRB), Stardust (Sdt), or MPP5 (Membrane Palmitoylated Protein 5) in mammals also known as PALS1 (protein associated with Lin-7 1) and PALS1-associated tight junction protein (PATJ); the lateral SCRIBBLE complex is made with Scribble (SCRIB), lethal giant larvae (LGL), and Disc large (DLG). These polarity complexes interact with and regulate each other

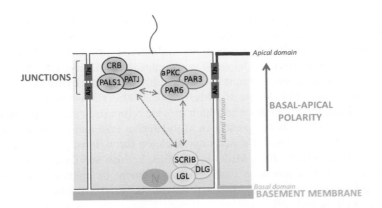

FIGURE 5.2 Localization and protein composition of the three polarity complexes needed to pattern cell polarity in epithelial cells: PAR, CRUMBS and SCRIBBLE according to what was described in Le Bivic (2013).

and are themselves regulated by signaling pathways in a complex manner (Assémat et al., 2008; Le Bivic, 2013).

5.1.6 Do All Metazoans Have (the Same) Epithelia?

There is no doubt that the last common ancestor of all extant animals was multicellular (King and Rokas, 2017). Independent of the variability in timing, relative importance and order of cell mechanisms involved in developing an epithelium among phyla, and sometimes among species of a same phylum (Ereskovsky, 2010), in metazoans, after the acquisition of multicellularity, the patterning of an epithelial or epithelial-like cell sheet is one of the earliest important events that takes place during embryogenesis, most often at the blastula stage. The acquisition of such an epithelial level of organization is intimately connected with the acquisition of cellular adhesion and cell-communication toolkits. One of the main questions raised is: did these key cell properties emerge once or several times? In other words, are all animal epithelia and epithelial-like sheets homologous among metazoans? And is it possible to trace back the origin and evolution of all animal epithelial cell types?

To try to answer this question, the first step is to examine the presence/absence in non-bilaterian taxa of the previously cited three histological criteria defining an animal epithelium (Figure 5.3) in order to: 1) decipher whether the term and definition of "epithelium" refers to equivalent structures throughout Metazoa and 2) determine, if epithelia are an ancestral character, which epithelial features were ancestrally present.

The first finding is that the organization of epithelial or epithelial-like sheets found in non-bilaterian phyla is much more variable—even within a single phylum—than it is in bilaterians. The second obvious finding is that, among non-bilaterians, only cnidarian epithelia have the three criteria of polarity, basement membrane, and junctions (whatever their type) at the same time (Ganot et al., 2015; Magie and Martindale,

Origin and Evolution of Epithelial Cell Types

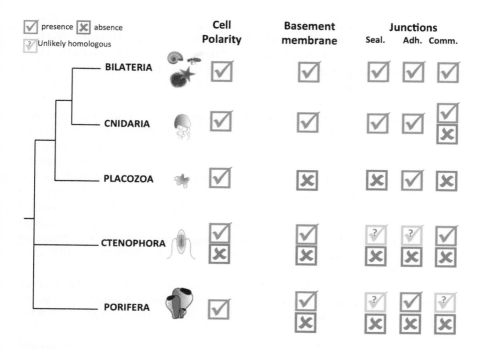

FIGURE 5.3 According to histological observations (Fidler et al., 2017; Ganot et al., 2015; Ledger, 1975; Leys and Riesgo, 2012; Magie and Martindale, 2008; Moroz and Kohn, 2016; Satterlie and Case, 1978; Smith and Reese, 2016) the joint presence of the three criteria supposed to define an epithelium in animals is far from being systematically conserved in non-bilaterian phyla. For junctions: "seal." stands for sealing, "adh." stands for adhesive and "comm." for communicating. In this schema, if two symbols are indicated for the same character in one phylum, it means that the observation can be different depending on the species considered. The status "unlikely homologous" means that a type of junction harboring histological features compatible with either sealing, adhesive, or communicating properties were described, but that their characteristics suggest that they are different from what is found typically in other animals (for sponges: [Adams et al., 2010; Leys and Hill, 2012; Leys and Riesgo, 2012; Leys et al., 2007, 2009]; for ctenophores: [Hernandez-Nicaise et al., 1989; Tamm and Tamm, 1991, 2002]). This schema shows that either many non-bilaterian species do not fit the definition of an epithelium or that the definition postulated decades ago does not depict and fit the diversity really found across animal epithelia.

2008). In contrast, Placozoa do not match the accepted definition of an epithelium (according to the very limited data available so far on this phylum) because of the absence of a basement membrane (Fidler et al., 2017; Ruthmann et al., 1986). As far as Ctenophora and Porifera are concerned, only some of the species have tissues that fully meet the definition of epithelium; this is the case for the Homoscleromorpha class in Porifera (Belahbib et al., 2018; Boute et al., 1996; Ereskovsky et al., 2009; Leys and Hill, 2012; Leys and Riesgo, 2012) and the genera *Beroe* and *Pleurobrachia* in Ctenophora (according to the limited data available so far on this phylum) (Fidler et al., 2017). It is therefore very surprising that the presence of a *bona fide* epithelium was questioned only for sponges. Indeed, while Porifera were classically excluded from the "Eumetazoa" clade partly because of the absence of "true" epithelia with a basement membrane, and even several years after Boute et al. (1996) showed a basement membrane existed in homoscleromorph sponges (Ereskovsky et al., 2009), in contrast, the term "epithelium" is currently used for Placozoa (Armon et al., 2018; Smith and Reese, 2016) and Ctenophora (Tamm and Tamm, 2002). Given such terminological inconsistency, there are two solutions: 1) the present definition of epithelium with three criteria is kept as it is, meaning that Placozoa, some Ctenophora, and most Porifera have to be considered devoid of epithelia; 2) or consider that the epithelium is a synapomorphy of metazoans with either the presence or absence of a basement membrane. This second solution has the positive outcome of taking into account the obvious conservation of physiological properties (mechanical resistance and occlusion of small molecules) that is found in epithelia of a demosponge (a group lacking basement membranes), a property that appears to be shared with bilaterian epithelia that have occluding junctions (Adams et al., 2010; Dunn et al., 2015; Leys and Riesgo, 2012).

Today, according to most authors, there is no longer any doubt about the presence of functional epithelia in the last common ancestor of animals that harbor cell polarity, adhesion properties, and basement membrane support (Fidler et al., 2017; King and Rokas, 2017; Leys and Hill, 2012; Leys and Riesgo, 2012; Medwig and Matus, 2017). We now have to explore how and when such an epithelium originated and evolved; this is the item of the following section.

Tracing back the evolution of epithelial cell types is not only related to the questionings concerning germ layer homology; we have also to consider that the similarity of cell features (here presence of junctions, polarity, and basement membrane, as previously mentioned) can result from convergent evolution by co-option of similar molecular players, or that shared proteins can have undergone neofunctionalization. In order to try to tackle this issue, the recent development of comparative whole organism single cell transcriptomic approaches (sc-RNAseq) should make it possible to define a cell type-specific core regulatory complex (CoRC) (Arendt, 2008; Arendt et al., 2016; Marioni and Arendt, 2017). Thanks to the conservation of regulatory mechanisms, a homologous cell type is expected to remain recognizable across species (Arendt, 2005, 2008). We will also discuss this point in section 2.

5.2 THE ORIGIN OF ANIMAL EPITHELIA

In order to understand the origin of this metazoan synapomorphy, it is useful to study the presence of "preadapted" molecular tools, those that were present before the emergence of animals. To do so, even if the data are scarce, comparative genomics of the closest unicellular relatives of metazoans are needed; choanoflagellates as the sister group of metazoans is, of course, of particular interest, but other members of Holozoa (such as Filasterea) and Opimoda (including Opisthokonta and Amoebozoa) can be useful (Ferrer-Bonet and Ruiz-Trillo, 2017; Olson, 2013; Richter and King, 2013; Richter et al., 2018; Sebé-Pedrós et al., 2017).

5.2.1 Some Genes Encoding Extracellular Matrix Proteins Predate the Emergence of Animal Epithelia

In bilaterians and cnidarians, epithelial morphogenesis has been shown to rely strongly on interactions between receptors on the epithelial sheet and components of the extracellular matrix (ECM), including the basement membrane when it is present (Aufschnaiter et al., 2011; Dzamba and DeSimone, 2018; Fidler et al., 2017; Sekiguchi and Yamada, 2018).

A few genes encoding important extracellular matrix proteins or their cellular receptors predate the emergence of Metazoa; indeed, genes encoding integrins and integrin adhesion machinery are present in unicellular relatives of metazoans (see section 2.2) (Abedin and King, 2010; Babonis and Martindale, 2017; Sebé-Pedrós and Ruiz-Trillo, 2010; Sebé-Pedrós et al., 2010; Suga et al., 2013). In both Filasterea (*C. owczarzaki*) and Choanoflagellata (*M. brevicolis* and *Salpingoeca rosetta*), several proteins, which contain protein domains that are present in animal ECM proteins (such as for example LAM G, EGF, fibronectin III) are predicted in the genomes of these organisms. But these protein domains are not combined in the same way as they are in Metazoa (Fairclough et al., 2013; King et al., 2008; Suga et al., 2013; Williams et al., 2014). This means that most of the proteins found in animal ECM and basement membrane (for instance the laminin and fibronectin key components) are metazoan innovations and emerged probably in part by domain shuffling (Babonis and Martindale, 2017; Fahey and Degnan, 2012; Richter et al., 2018; Suga et al., 2013; Williams et al., 2014). Interestingly, most recent studies suggest that a canonical type IV collagen could have been already present in the common ancestor of Filozoa (Fidler et al., 2018; Grau-Bové et al., 2017) therefore suggesting a more ancient origin of type IV collagen than previously accepted and its neo-functionalization in metazoans, as it has also been suggested for integrin adhesome or cadherins (King et al., 2008; Sebé-Pedrós and Ruiz-Trillo, 2010; Sebé-Pedrós et al., 2010). In other words, the characterization of type IV collagen in non-metazoans challenges the previously accepted idea that spongin short chain collagen (SSCC) is ancestral to type IV collagen (Aouacheria et al., 2006; Leys and Riesgo, 2012).

5.2.2 Integrins: An Ancient Protein Family

As previously mentioned, (section 1.4), the key proteins shared by focal adhesions (FAs) and hemidesmosomes (HDs) are integrins. Until 2010, because genes encoding integrins were not retrieved in plant, fungi, and choanoflagellate genome surveys, integrins were thought to be a metazoan innovation. However, genomic data from several unicellular opisthokont species show that integrins, as well as components of the mechanism by which integrins interact with cells (Focal Adhesion kinase and non-receptor protein kinase Src), were present in the last common ancestor of Filasterea and Metazoa (Sebé-Pedrós and Ruiz-Trillo, 2010; Sebé-Pedrós et al., 2010). In addition, both integrin α and β in these premetazoan taxa possess the amino acid motifs in their cytoplasmic tails involved in interactions with intracellular scaffolding and signaling proteins, therefore suggesting that they could interact in a manner similar to their metazoan homologs. Though the role of the integrins in unicellular eukaryotes is currently unknown, it has been shown very recently that the *Capsaspora* integrins are able to recruit human talin. Therefore, this result suggests that the regulation of integrins via talin activity is conserved in Filozoa and predates the emergence of Metazoa (Baade et al., 2019).

On the other hand, to build HDs, plectin is required to form bridges between the cytoplasmic keratin intermediate filament network and the integrins. To date, no plectin has been characterized so far in non-metazoan lineages, but other linker proteins allowing the binding between integrins and the cytoskeleton are common among Opimoda (Sebé-Pedrós et al., 2010).

However, integrins predate the emergence of animals and may play an ancestral role in signaling, and their role in cell-ECM adhesion might have been co-opted in metazoans (Sebé-Pedrós and Ruiz-Trillo, 2010; Sebé-Pedrós et al., 2010).

5.2.3 GENES ENCODING CELL-CELL ADHESIVE PROTEINS PREDATE THE EMERGENCE OF ANIMAL EPITHELIA

In section 1.4, we described the types of junctions found in bilaterians and their molecular composition. To date, because actin-based adhesive-like junctions (Ganot et al., 2015; Magie and Martindale, 2008; Tamm and Tamm, 1987) and adhaerens junctions (Ereskovsky et al., 2009; Fahey and Degnan, 2010; Leys and Riesgo, 2012; Leys et al., 2009) were described in Ctenophora and Porifera respectively, it was assumed that adhaerens junctions (AJs) constituted the ancestral type of junctions of animals. As a consequence, the cadherins and catenins involved in the establishment of AJs (section 1.4) were one of the first structures sought in unicellular relatives of animals.

According to the still somewhat scarce genomic data available, it seems that at least three cadherin families were present in the last common ancestor of Choanozoa (taxa grouping Choanoflagellata and Metazoa): lefftyrins, coherins, and hedglings (Nichols et al., 2012). Among these families, some were lost secondarily in bilaterians or earlier. To date, only one cadherin gene was found in the filasterean *C. owczarzaki* (King et al., 2008; Nichols et al., 2012; Richter and King, 2013). It is therefore now obvious that cadherins predate the origin of metazoans. But only metazoans have *bona fide* classical cadherins containing the two domains required for interaction with catenins (juxtamembrane domain, JMD, and the catenin-binding domain, CBD) (Clarke et al., 2016; Miller et al., 2013; Murray and Zaidel-Bar, 2014).

As far as catenins are concerned, the amoebozoan *Dictyostelium discoideum* possesses catenins (Aardvark and Ddα-catenin), which are involved in establishing epithelial-like sheets (Dickinson et al., 2011, 2012). Interestingly, Aardvark-related proteins are also present in choanoflagellates and filastereans (Nichols et al., 2012; Suga et al., 2013). Aardvark and Ddα-catenin are considered to be close relatives of beta- and alpha- catenins because of shared domain features (Miller et al., 2013, 2018; Richter et al., 2018; Suga et al., 2013). As a consequence, it can be assumed that the joint presence of cadherins and catenin-related proteins predate the emergence of

Metazoa even if it is predicted they do not interact in non-metazoans. Unfortunately, there is no experimental data so far available to establish the function of these proteins in these organisms (Nichols et al., 2012). But the recent acquisition of a transfection protocol in choanoflagellates is expected to enable to soon fill this gap (Booth et al., 2018).

Concerning genes encoding proteins involved in other types of junctions, none of the components of gap junctions have been characterized in non-metazoans yet (Moroz and Kohn, 2016), while, in contrast, a claudin-like gene was retrieved in *Capsaspora* and *Monosiga* genomes (Ganot et al., 2015) (see Figure 5.5).

5.2.4 Do Some Genes Encoding Members of Cell Polarity Complexes Predate the Emergence of Animal Epithelia?

Though the typical collar cell of choanoflagellates shows an obvious basal-apical cell polarity, to date the genes encoding the three metazoan polarity complexes (CRUMBS, SCRIBBLE, and PAR, Figure 5.2) remain nearly unexplored. Among the various players involved in polarity complexes (see section 1.5), only DLG, a member of the SCRIBBLE complex was shown to predate the emergence of the metazoans and has been found in Choanoflagellata, Filasterea, and Ichthyosporea (Belahbib et al., 2018; Fahey and Degnan, 2010; Ganot et al., 2015; Le Bivic, 2013; Murray and Zaidel-Bar, 2014; Richter et al., 2018). For now, without further evidence to the contrary, all other components are considered metazoan innovations. In addition, the key pathways known to regulate these three epithelial polarity complexes (among which are the Wnt and PCP pathways) are also considered to be metazoan specific (Babonis and Martindale, 2017).

If metazoans evolved these specific means to control their epithelial cell polarity, it is obvious that cell polarity is more generally necessary for proper functioning of numerous other cell types such as neurons in animals, pollen tubes in plants, and hyphae in Fungi. The achievement of cell polarity is intimately linked to cytokinesis, orientation of mitotic spindles, and asymmetric separation. Some molecular elements controlling cell polarity are known to be highly conserved across eukaryotes, including animals, in epithelial and non-epithelial cell types, such as MOB proteins and Rho GTPase (Bornens, 2018; Hoff, 2014; Lefèbvre et al., 2012; Slabodnick et al., 2014).

5.2.5 Main Conclusion: Cell Adhesion and Polarity Emerged Several Times during the Evolution of Holozoa

According to the previously cited results in Choanoflagellata and Filasterea, the proteins involved in cell-cell adhesion and cell-matrix adhesion to achieve aggregative/transient multicellular stages in these organisms are different from those found in metazoans. Nevertheless, key protein domains involved in major cell-cell or cell-matrix protein interactions are highly conserved in holozoans. This means that, during the evolution of holozoans, aggregative/multicellular stages probably emerged thanks to intense domain shuffling of shared building blocks, as Lego® bricks, and duplication-divergence events in already present gene families (e.g. cadherins and catenins). Thanks to these ancestral domains and families, the metazoan lineage evolved specific tools, representing metazoan synapomorphies (such as polarity complexes, classical E-cadherin, and signaling pathways), to achieve permanent multicellular levels of organization with epithelial features.

5.3 THE PROBABLY ANCESTRAL FEATURES OF THE METAZOAN EPITHELIUM

5.3.1 THE COMPOSITION OF ANCESTRAL BASEMENT MEMBRANE

To date, comprehensive biochemical experiments to determine the protein composition of the basement membranes of animals have only been carried out in cnidarians and bilaterians (Fidler et al., 2014, 2018; Halfter et al., 2015). As a consequence, only partial biochemical information is available for the other animal lineages, particularly for sponges and ctenophores. Most of our present knowledge for these two phyla therefore relies on indirect hypotheses based on transcriptomic and genomic surveys.

A few studies have compared the extracellular matrix (ECM) genes found in the genomes of sponges and cnidarians with that conserved across bilaterians. These comparisons revealed the presence of an ancestral core set of ECM components, receptors, and degrading proteases (Adams, 2013, 2018; Özbek et al., 2010; Tucker and Adams, 2014; Williams et al., 2014). Given what is known about bilaterians and cnidarians, the extracellular matrix is mainly composed of type I collagen, elastin, and fibronectins, but there is also a great variety of glycoproteins (such as laminin), proteoglycans, and different types of collagens (except type IV). Its density, fluidity, and composition is variable between taxa and between different tissues of the same taxa (Fidler et al., 2018; Hynes, 2012; Kular et al., 2014).

In contrast, the composition of the basement membrane seems less variable between taxa (even though slight variation can occur between tissues). Indeed, in all cases, the main components are laminins and type IV collagen, interconnected with mainly nidogen and perlecan, sometimes with additional molecules such as agrin (Fidler et al., 2017, 2018; Medwig and Matus, 2017; Pozzi et al., 2017; Sekiguchi and Yamada, 2018). According to genome surveys, only genes encoding type IV collagen and laminin are present in ctenophores and sponges. Surprisingly, despite of the absence of the connecting proteins nidogen and perlecan, they are able to build a basement membrane in some species (at least, this is true of *Beroe ovata* and *Pleurobrachia pileus* among Ctenophores and of Homoscleromorpha among sponges) (Fidler et al., 2017, 2018; Kenny et al., 2020). This finding suggests that these two proteins (type IV collagen and laminin) are essential and shared components of BM in animals and therefore constitute the core toolkit already present in the last common ancestor of all extant metazoans. This also means that the other BM components (perlecan and nidogen) emerged more recently during metazoan evolution. Future experiments are needed to understand the differences in the organization and mechanical properties of the BM between cnidarians/bilaterians and ctenophores/homoscleromorph sponges.

These recent findings in ctenophores and sponges also mean that the absence of a BM in some ctenophores, such as *Mnemiopsis leidyi*, and in the placozoan *Trichoplax adhaerens*, or of genes encoding laminin and type IV collagen (as in some demosponges for instance) are secondary losses. Interestingly, in contrast, the calcareous sponge *Sycon coactum* possesses genes encoding several BM proteins whereas a BM structure has never been observed in this sponge class (Leys and Riesgo, 2012; Riesgo et al., 2014). More in-depth domain analyses and functional experiments have to be performed to fully understand this apparent discrepancy between gene content and tissue structure (discussed in Renard et al., 2018).

5.3.2 THE ANCESTRAL TOOLKIT TO ACHIEVE CELL POLARITY DEPENDS ON NON-BILATERIAN RELATIONSHIPS

Cell polarity is an ancestral feature of animal epithelia (Figure 5.3). It now remains to be determined whether or not the molecular toolkit involved in the establishment of coordinated cell polarity is the same in non-bilaterian animals as in bilaterians. According to recent gene surveys (Belahbib et al., 2018; Riesgo et al., 2014), all the genes encoding the nine proteins involved in PAR, CRUMBS, and SCRIBBLE polarity complexes are present in Cnidaria, Placozoa, and Porifera, whereas the whole set of genes needed to build a CRUMBS complex is absent in Ctenophora, as well as the *Scribble* gene needed to complete the SCRIBBLE complex (Table 5.2, Figures 5.4 and 5.5).

TABLE 5.2
Presence (x)/Absence(0) in the Four Non-Bilaterian Phyla (Porifera, Ctenophora, Placozoa, and Cnidaria) of Genes Encoding Proteins Involved in the Three Cell Polarity Complexes, Namely the CRUMBS Complex, the SCRIBBLE Complex, and the PAR Complex (Figure 5.2), According to Transcriptomic and Genomic Surveys (Belahbib et al., 2018; Le Bivic, 2013; Riesgo et al., 2014)

		Porifera	Ctenophora	Placozoa	Cnidaria
PAR	PAR 3	x	x	x	x
	PAR 6	x	x	x	x
	aPKC	x	x	x	x
CRUMBS	CRB	x	0	x	x
	PAT J	x	0	x	x
	PALS1	x	0	x	x
SCRIBBLE	SCRIB	x	0	x	x
	LGL	x	x	x	x
	DLG	x	x	x	x

Because the relative phylogenetic position of Ctenophora and Porifera is still uncertain (Simion et al., 2017; Whelan et al., 2017; Kapli and Telford, 2020; Francis and Canfield, 2020; reviewed in King and Rokas, 2017 and in Schenkelaars et al., 2019) two scenarios are possible: either the nine polarity genes (DLG predates the emergence of Metazoa) were present in the common metazoan ancestor and then some were lost secondarily in ctenophores, or only five of these genes were ancestral and the other four genes would represent innovations acquired later during animal evolution (Figure 5.4). Moreover, whatever the scenario considered, there is not yet any experiment establishing the function of these proteins and their ability to form complexes (even if residue and domain analyzes tend to support this hypothesis, see Belahbib et al., 2018). Interestingly, a very recent immunolocalization of the protein PAR6 in the ctenophore *Mnemiopsis leidyi* may suggest the functional conservation of this protein outside the cnidarian-bilaterian lineage (Salinas-Saavedra and Martindale, 2019).

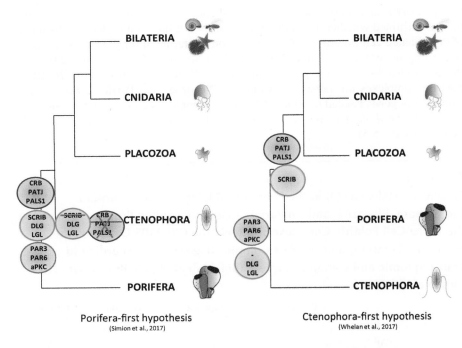

FIGURE 5.4 Two scenarios for evolution of the three polarity complexes (CRUMBS, PAR, and SCRIBBLE complexes) taking into account the uncertainty of phylogenetic relationships at the base of the animal tree (Porifera or Ctenophora as sister group of all other metazoans): either the 3 polarity complexes were already present in the last common ancestor of all extant animals or partially two of them were present. The most recent phylogenomic study (Kapli and Telford, 2020) supports the first scenario.

Origin and Evolution of Epithelial Cell Types

5.3.3 Were Adhesive Junctions Present Ancestrally?

According to the analyses of the different types of junctions present in bilaterians and cnidarians, it is clear that their common ancestor must have had AJs, gap junctions, and SJs, but the origin of these major innovations remains unknown.

5.3.3.1 Communicating Junctions

In bilaterians, gap junctions are either composed of innexin (invertebrates) or connexin (vertebrates). Whatever the evolutionary origin of these two proteins (homology vs. convergence), the gap junctions they compose show amazing structural and functional similarities (Skerrett and Williams, 2017). In cnidarians, innexin also composes gap junctions (Alexopoulos et al., 2004; Takaku et al., 2014), which means that innexin-gap-junctions were already present in the last ancestor of bilaterians and cnidarians. Gap junctions have been reported in ctenophores (Satterlie and Case, 1978), and interestingly, a gene encoding innexin has been characterized in *Pleurobrachia* (Moroz and Kohn, 2016). Nevertheless, given that innexin can participate in cell-cell communication via the establishment of either gap junctions or hemichannels (Güiza et al., 2018), the localization of the innexin protein at the level of the gap junction is needed to establish the possible functional conservation of innexin in ctenophores. In contrast, until now, neither the gap junction structure nor the innexin genes have been found in either sponges or placozoans (Moroz and Kohn, 2016; Smith and Reese, 2016). A very different type of communicating junction, called a plug junction, which is clearly unrelated to gap junctions found in other animals, was described in glass sponges (Hexactinellida) (Leys et al., 2006, 2007). Therefore, if we assume 1) that the innexin protein is involved in the formation of gap junctions in ctenophores and 2) that the Ctenophora-first hypothesis is correct, then gap junctions were present ancestrally in animals and lost secondarily in sponges and placozoans. Because none of these hypotheses have so far been validated, it seems premature to draw such a conclusion.

5.3.3.2 Sealing/Occluding Junctions

As far as sealing junctions are concerned, septate junctions are present in Cnidaria. Despite ultrastructural variations observed by electron microscopy and incomplete information about their molecular composition, bilaterian and cnidarian SJs are currently considered to be evolutionary related (Ganot et al., 2015). Again, as for the other structures examined in this chapter, our knowledge from the three other non-bilaterian phyla is too limited, which prevents us from outlining a clear evolutionary scenario for the origin of SJs. In Placozoa, ladder-like structures reminiscent of SJs have been described (Ruthmann et al., 1986), but more precise observations are needed to conclude whether or not these structures are *bona fide* SJs (Ganot et al., 2015; Smith and Reese, 2016). In ctenophores, no SJs have been reported so far, although "atypical" junctional structures (without septa) have been characterized (Hernandez-Nicaise et al., 1989; Magie and Martindale, 2008). Although sponge epithelia clearly have sealing properties (Adams et al., 2010), SJs have been reported in only one study from the calcisponge *Sycon ciliatum*; later studies on the same genus (but a different species) have not so far confirmed this feature, although sclerocytes, where the SJs were reported to occur, were not studied precisely (Eerkes-Medrano and Leys, 2006; Ledger, 1975; Leys et al., 2009). Interestingly, despite the absence of unquestionable SJs outside the Cnidaria-Bilateria lineage, genes encoding core components of SJs are present in placozoans and some in sponges and ctenophores (Ganot et al., 2015; Riesgo et al., 2014). Therefore, four proteins (claudin-like, neuroglian, contactin, neurexin IV), which are considered major structural components of SJs, would have predated the emergence of the Bilateria-Cnidaria lineage (Figure 5.5). Considering the incongruences between the different analyses performed so far in ctenophores and sponges, it is currently impossible to determine whether these genes were present ancestrally or not (Chapman et al., 2010; Fahey and Degnan, 2010; Ganot et al., 2015; Leys and Riesgo, 2012; Riesgo et al., 2014; Suga et al., 2013).

Thus, much remains to be explored in order to 1) understand how these three non-bilaterian phyla, namely Placozoa, Ctenophora, and Porifera, achieve the sealing of their epithelium, 2) establish the roles of retrieved genes, and 3) trace back the origin of bilaterian-cnidarian septate junctions.

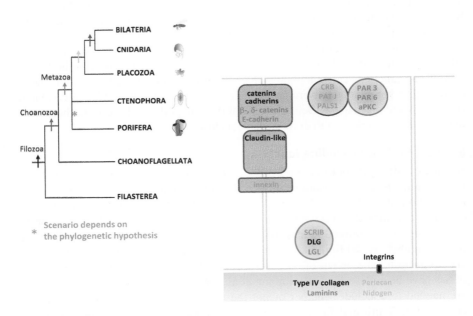

FIGURE 5.5 Diagram summarizing the emergence of genes involved in key epithelial features (junctions, basement membrane, and polarity complexes) according to present available data. In the figure, proteins were located in order to depict their function in cnidarians and bilaterians, even though there is no evidence of the functional conservation of these proteins outside the cnidarian-bilaterian lineage.

5.3.3.3 Adhesive Junctions

In contrast to the two previous types of junctions, because of the clear description of adhaerens junctions or adhesive belts in some sponges (Homoscleromorpha and Calcarea), placozoans, and ctenophores (Ereskovsky et al., 2009; Fahey and Degnan, 2010; Ganot et al., 2015; Leys and Riesgo, 2012; Leys et al., 2009; Magie and Martindale, 2008; Ruthmann et al., 1986; Smith and Reese, 2016; Smith et al., 2014; Tamm and Tamm, 1987), until recently, it was taken for granted that adhesive junctions represented the most ancestral type of junctions. Nevertheless, recent transcriptomic and genomic surveys (Belahbib et al., 2018; Riesgo et al., 2014) and deep analyses of key functional domain and residues (Belahbib et al., 2018) started casting doubt on this common view. Indeed, whereas only two classes of sponges harbor adhaerens junctions (Homoscleromorpha and Calcarea), all four sponge classes possess the genes coding for the four proteins involved in adhaerens junctions, namely E-cadherin, alpha-, beta-, and delta-catenins (CCC complex, Figure 5.1). Moreover, the key domains of the E-cadherin needed to establish the CCC complex (GBM and JMD) are well conserved in three of these classes (while divergent in glass sponges) (Belahbib et al., 2018), and the first recent biochemical experiments conducted in a demosponge (which does not have adhaerens junctions) suggest that this canonical complex is formed *in vitro* (Schippers et al., 2018). Despite the fact that these new data are very exciting and represent an undeniable step forward for sponge biology, they unfortunately fail to explain the functional role of the CCC complex in sponge classes that lack adhaerens junction-like structures. It will be important to explore the homology of adhaerens junctions of homoscleromorph sponges and of cnidarians/bilaterians. Recently, the involvement of vinculin in cell-cell and cell-ECM contacts in *Oscarella pearsei* has been proposed following its tissue localization by immunolocalization (Miller et al., 2018). This is highly reminiscent of what is known in other animals (Carisey and Ballestrem, 2011; Carisey et al., 2013; Miller et al., 2018), but until E-cadherin is localized in this class of sponge, we will not be able to make the link between the structure (AJs) and its protein composition.

The most surprising result of the analyses conducted by Belahbib et al. (2018) was the highly divergent key cytoplasmic domains of the E-cadherin of ctenophores, which prevent predictions of a possible interaction of this protein with catenins, and casts doubt on the formation of a CCC complex in this phylum. Again, as for sponges, biochemical and functional experiments are needed to confirm/infirm *in vitro* and *in vivo* these *in silico* predictions. If this is confirmed, and depending on the phylogenetic hypothesis considered, this may challenge the idea that the last common ancestor of animals was already able to form CCC complexes involved in the formation of adhesive junctions. Indeed the possibility of neofunctionalization of cadherins and catenins is well illustrated by the case of the emergence of desmosomes in vertebrates (for review see Green et al., 2020).

5.3.4 Epithelial Regulatory Signatures

In recent years, sc-RNAseq has been used to compare similar cell types across different species. This approach is expected to provide insight into the evolution of cellular lineages (reviewed in Marioni and Arendt, 2017). The general principle of this approach is that: 1) a cell type is defined by a unique combination of transcription factors, the terminal selectors, that form a core regulatory complex (CoRC) needed to regulate cell type–specific effector genes (Arendt et al., 2016; Hobert et al., 2010); 2) evolutionary relatedness between cell types should be visible through the sharing of similarities of CoRCs and effectors.

So far, no study has focused on the evolution of epithelial cell types. Nevertheless, a few transcription factors seem to be highly conserved in some epithelial cells across the animal kingdom and may be expected to be part of the CoRC needed to regulate the cell fate of at least some epithelial cell types. This is, for example, the case of epithelial specific Ets factors (ESE) that regulate epithelial processes, such as epithelial proliferation and differentiation, in different mammalian epithelial cell types, notably by modulating terminal differentiation pathways (Feldman et al., 2003) and that also seem to be specific markers of one of the epithelial layers found in sponges (pinacocytes forming the pinacoderm layer) (Sebé-Pedrós et al., 2018). In the same way, the transcription factor *Grainyhead* that may play a role in regulating domain-specific effector genes for some epithelial cell types (Achim et al., 2018; Boglev et al., 2011) has been found recently to be expressed specifically in some epithelial cell types in the ctenophore *Mnemiopsis leidyi* and the choanoderm (formed by choanocytes) of the sponge *Amphimedon queenslandica* (Sebé-Pedrós et al., 2018).

Present data are obviously insufficient to decipher evolutionary relationships between epithelial cell types among animal lineages, but ongoing projects are expected to provide clues in future years. The main difficulties to deal with at such a large evolutionary scale are: 1) establishing the precise orthology and paralogy relationships of all expressed genes (Altenhoff et al., 2019), 2) acquiring high-quality reference genomes and gene annotation for non-bilaterian lineages which is currently far from accomplished (Renard et al., 2018), 3) attaining a better knowledge of cell types, especially in non-bilaterian phyla (Sebé-Pedrós et al., 2018), and of their ability to transdifferentiate (Sogabe et al., 2016).

5.4 CONCLUSIONS AND FUTURE CHALLENGES

It is now clear that the main features of bilaterian epithelia predate the emergence of the bilaterian lineage; indeed, all these features are clearly shared with cnidarians, meaning that the last common ancestor of cnidarians and bilaterians already had an epithelium with cell polarity, basement membrane, and three types of junctions, and that these features are achieved with the same molecular toolkit. To trace back the origin of these features, it is therefore necessary to look at the three other non-bilaterian phyla, namely sponges, ctenophores, and placozoans. Unfortunately, the knowledge available on these three phyla mainly relies 1) on classical electron microscopy and 2) on gene surveys. These data describe major discrepancies between gene content and observed histological features. The other great lesson learned in the last years is that even though the basement membrane was probably present ancestrally, it was lost several times during animal evolution. This finding challenges the common dogma on the absolute requirement of the presence of this structure to compose and pattern an epithelium. Indeed, it has been shown in demosponges that the achievement of epithelial properties can be reached without the presence of a basement membrane (Adams et al., 2010). Nevertheless, the different regeneration processes found in different sponge species (Borisenko et al., 2015; Ereskovsky et al., 2015) suggest that the presence or absence of a basement membrane may influence the dynamics of epithelial morphogenetic processes. Much remains to be explored from a functional point of view in order to understand the consequences of secondary losses of the basement membrane in several species.

The present state of this exciting quest is that biochemical and functional experiments are needed to establish the exact role of the proteins in these three non-bilaterian lineages. Recent papers show that there are promising growing efforts in this direction. Only these proteomic and functional data will enable us to determine whether the basement membrane and the cell polarity already present in the last common ancestor of all extant animal relied on the same molecular players and determine which types of junction were present ancestrally.

ACKNOWLEDGEMENTS

The authors are grateful to the AM*Idex foundation, Excellence Initiative of Aix-Marseille Université, that funded a first project that enabled the study of epithelia in the sponge *Oscarella lobularis* (n° ANR-11-IDEX-0001-02) and recently a second international project that now enables us to begin comparative scRNAseq studies to trace back epithelial cell type evolution (AMX-18-INT-021). The authors also acknowledge the CNRS for the funding of the French-Australian PICS project STraS funding the development of cell staining protocols in order to study epithelia dynamics.

The authors also thank Sally Leys for proofreading the text.

REFERENCES

Abedin, M., and King, N. (2010). Diverse evolutionary paths to cell adhesion. *Trends Cell Biol.* 20, 734–742.

Achim, K., Eling, N., Vergara, H.M., Bertucci, P.Y., Musser, J., Vopalensky, P., Brunet, T., Collier, P., Benes, V., Marioni, J.C., et al. (2018). Whole-body single-cell sequencing reveals transcriptional domains in the annelid larval body. *Mol. Biol. Evol.* 35, 1047–1062.

Adams, J.C. (2013). Extracellular matrix evolution: An overview. In *Evolution of Extracellular Matrix*, F.W. Keeley and R.P. Mecham, eds. (Berlin, Heidelberg: Springer Berlin Heidelberg), pp. 1–25.

Adams, J.C. (2018). Matricellular proteins: Functional insights from non-mammalian animal models. *Curr. Top. Dev. Biol.* 130, 39–105.

Adams, E.D.M., Goss, G.G., and Leys, S.P. (2010). Freshwater sponges have functional, sealing epithelia with high transepithelial resistance and negative transepithelial potential. *PLoS One* 5, e15040.

Alexopoulos, H., Böttger, A., Fischer, S., Levin, A., Wolf, A., Fujisawa, T., Hayakawa, S., Gojobori, T., Davies, J.A., David, C.N., et al. (2004). Evolution of gap junctions: The missing link? *Curr. Biol.* 14, R879–R880.

Altenhoff, A.M., Glover, N.M., and Dessimoz, C. (2019). Inferring orthology and paralogy. In *Evolutionary Genomics: Statistical and Computational Methods*, M. Anisimova, ed. (New York, NY: Springer New York), pp. 149–175.

Aouacheria, A., Geourjon, C., Aghajari, N., Navratil, V., Deléage, G., Lethias, C., and Exposito, J.-Y. (2006). Insights into early extracellular matrix evolution: Spongin short chain collagen-related proteins are homologous to basement membrane type IV collagens and form a novel family widely distributed in invertebrates. *Mol. Biol. Evol. 23*, 2288–2302.

Arendt, D. (2005). Genes and homology in nervous system evolution: Comparing gene functions, expression patterns, and cell type molecular fingerprints. *Theory Biosci. 124*, 185–197.

Arendt, D. (2008). The evolution of cell types in animals: Emerging principles from molecular studies. *Nat. Rev. Genet. 9*, 868–882.

Arendt, D., Musser, J.M., Baker, C.V.H., Bergman, A., Cepko, C., Erwin, D.H., Pavlicev, M., Schlosser, G., Widder, S., Laubichler, M.D., et al. (2016). The origin and evolution of cell types. *Nat. Rev. Genet. 17*, 744–757.

Armon, S., Bull, M.S., Aranda-Diaz, A., and Prakash, M. (2018). Ultrafast epithelial contractions provide insights into contraction speed limits and tissue integrity. *PNAS 115*, E10333–E10341.

Assémat, E., Bazellières, E., Pallesi-Pocachard, E., Le Bivic, A., and Massey-Harroche, D. (2008). Polarity complex proteins. *Biochim. Biophys. Acta (BBA) Biomembr. 1778*, 614–630.

Aufschnaiter, R., Zamir, E.A., Little, C.D., Özbek, S., Münder, S., David, C.N., Li, L., Sarras, M.P., and Zhang, X. (2011). In vivo imaging of basement membrane movement: ECM patterning shapes Hydra polyps. *J. Cell Sci. 124*, 4027–4038.

Baade, T., Paone, C., Baldrich, A., and Hauck, C.R. (2019). Clustering of integrin β cytoplasmic domains triggers nascent adhesion formation and reveals a protozoan origin of the integrin-talin interaction. *Sci. Rep. 9*, 5728.

Babonis, L.S., and Martindale, M.Q. (2017). Phylogenetic evidence for the modular evolution of metazoan signalling pathways. *Philos. Trans. R. Soc. Lond. B Biol. Sci. 372*, 20150477.

Belahbib, H., Renard, E., Santini, S., Jourda, C., Claverie, J.-M., Borchiellini, C., and Le Bivic, A. (2018). New genomic data and analyses challenge the traditional vision of animal epithelium evolution. *BMC Genom. 19*, 393.

Boglev, Y., Wilanowski, T., Caddy, J., Parekh, V., Auden, A., Darido, C., Hislop, N.R., Cangkrama, M., Ting, S.B., and Jane, S.M. (2011). The unique and cooperative roles of the Grainy head-like transcription factors in epidermal development reflect unexpected target gene specificity. *Dev. Biol. 349*, 512–522.

Booth, D.S., Szmidt-Middleton, H., and King, N. (2018). Choanoflagellate transfection illuminates their cell biology and the ancestry of animal septins. *Mol. Biol. Cell 29*, 3026–3038. doi:10.1091/mbcE18080514.

Borisenko, I.E., Adamska, M., Tokina, D.B., and Ereskovsky, A.V. (2015). Transdifferentiation is a driving force of regeneration in *Halisarca dujardini* (Demospongiae, Porifera). *PeerJ 3*, e1211.

Bornens, M. (2018). Cell polarity: Having and making sense of direction—On the evolutionary significance of the primary cilium/centrosome organ in Metazoa. *Open Biol. 8*, 180052.

Boute, N., Exposito, J.-Y., Boury-Esnault, N., Vacelet, J., Noro, N., Miyazaki, K., Yoshizato, K., and Garrone, R. (1996). Type IV collagen in sponges, the missing link in basement membrane ubiquity*. *Biol. Cell 88*, 37–44.

Carisey, A., and Ballestrem, C. (2011). Vinculin, an adapter protein in control of cell adhesion signalling. *Eur. J. Cell Biol. 90*, 157–163.

Carisey, A., Tsang, R., Greiner, A.M., Nijenhuis, N., Heath, N., Nazgiewicz, A., Kemkemer, R., Derby, B., Spatz, J., and Ballestrem, C. (2013). Vinculin regulates the recruitment and release of core focal adhesion proteins in a force-dependent manner. *Curr. Biol.* 23, 271–281.

Chapman, J.A., Kirkness, E.F., Simakov, O., Hampson, S.E., Mitros, T., Weinmaier, T., Rattei, T., Balasubramanian, P.G., Borman, J., Busam, D., et al. (2010). The dynamic genome of Hydra. *Nature 464*, 592–596.

Clarke, D.N., Miller, P.W., Lowe, C.J., Weis, W.I., and Nelson, W.J. (2016). Characterization of the cadherin?Catenin complex of the sea anemone *Nematostella vectensis* and implications for the evolution of metazoan cell? Cell adhesion. *Mol. Biol. Evol. 33*, 2016–2029.

De Pascalis, C., and Etienne-Manneville, S. (2017). Single and collective cell migration: The mechanics of adhesions. *Mol. Biol. Cell 28*, 1833–1846.

Dickinson, D.J., Nelson, W.J., and Weis, W.I. (2011). A polarized epithelium organized by beta- and alpha-catenin predates cadherin and metazoan origins. *Science 331*, 1336–1339.

Dickinson, D.J., Nelson, W.J., and Weis, W.I. (2012). An epithelial tissue in Dictyostelium challenges the traditional origin of metazoan multicellularity. *Bioessays 34*, 833–840.

Dunn, C.W., Leys, S.P., and Haddock, S.H.D. (2015). The hidden biology of sponges and ctenophores. *Trends Ecol. Evol. 30*, 282–291.

Dzamba, B.J., and DeSimone, D.W. (2018). Extracellular matrix (ECM) and the sculpting of embryonic tissues. *Curr. Top. Dev. Biol. 130*, 245–274.

Eerkes-Medrano, D.I., and Leys, S.P. (2006). Ultrastructure and embryonic development of a syconoid calcareous sponge. *Invertebr. Biol. 125*, 177–194.

Ereskovsky, A. (2010). *The Comparative Embryology of Sponges*. Berlin: Springer, pp. 209–230.

Ereskovsky, A.V., Borchiellini, C., Gazave, E., Ivanisevic, J., Lapébie, P., Perez, T., Renard, E., and Vacelet, J. (2009). The Homoscleromorph sponge Oscarellalobularis, a promising sponge model in evolutionary and developmental biology. *Bioessays 31*, 89–97.

Ereskovsky, A.V., Borisenko, I.E., Lapébie, P., Gazave, E., Tokina, D.B., and Borchiellini, C. (2015). *Oscarella lobularis* (Homoscleromorpha, Porifera) regeneration: Epithelial morphogenesis and metaplasia. *PLoS One 10*: e0134566.

Fahey, B., and Degnan, B.M. (2010). Origin of animal epithelia: insights from the sponge genome: Evolution of epithelia. *Evol. Dev. 12*, 601–617.

Fahey, B., and Degnan, B.M. (2012). Origin and evolution of laminin gene family diversity. *Mol. Biol. Evol. 29*, 1823–1836.

Fairclough, S.R., Chen, Z., Kramer, E., Zeng, Q., Young, S., Robertson, H.M., Begovic, E., Richter, D.J., Russ, C., Westbrook, M.J., et al. (2013). Premetazoan genome evolution and the regulation of cell differentiation in the choanoflagellate *Salpingoeca rosetta*. *Genome Biol. 14*, R15.

Feldman, R.J., Sementchenko, V.I., and Watson, D.K. (2003). The epithelial-specific Ets factors occupy a unique position in defining epithelial proliferation, differentiation and carcinogenesis. *Anticancer Res. 23*, 2125–2131.

Ferrer-Bonet, M., and Ruiz-Trillo, I. (2017). Capsaspora owczarzaki. *Curr. Biol. 27*, R829–R830.

Fidler, A.L., Vanacore, R.M., Chetyrkin, S.V., Pedchenko, V.K., Bhave, G., Yin, V.P., Stothers, C.L., Rose, K.L., McDonald, W.H., Clark, T.A., et al. (2014). A unique covalent bond in basement membrane is a primordial innovation for tissue evolution. *Proc. Natl. Acad. Sci. USA. 111*, 331–336.

Fidler, A.L., Darris, C.E., Chetyrkin, S.V., Pedchenko, V.K., Boudko, S.P., Brown, K.L., Gray Jerome, W., Hudson, J.K., Rokas, A., and Hudson, B.G. (2017). Collagen IV and basement membrane at the evolutionary dawn of metazoan tissues. *Elife 6*, e24176.

Fidler, A.L., Boudko, S.P., Rokas, A., and Hudson, B.G. (2018). The triple helix of collagens – An ancient protein structure that enabled animal multicellularity and tissue evolution. *J. Cell Sci. 131*, jcs203950.

Fischer, R.S., Lam, P.-Y., Huttenlocher, A., and Waterman, C.M. (2019). Filopodia and focal adhesions: An integrated system driving branching morphogenesis in neuronal pathfinding and angiogenesis. *Dev. Biol. 451*, 86–95.

Francis, W.R., Canfield D.E. (2020). Very few sites can reshape the inferred phylogenetic tree. *PeerJ 8*, e8865.

Ganot, P., Zoccola, D., Tambutté, E., Voolstra, C.R., Aranda, M., Allemand, D., and Tambutté, S. (2015). Structural molecular components of septate junctions in cnidarians point to the origin of epithelial junctions in eukaryotes. *Mol. Biol. Evol. 32*, 44–62.

Garcia, M.A., Nelson, W.J., and Chavez, N. (2018). Cell-cell junctions organize structural and signaling networks to regulate epithelial tissue homeostasis. *Cold Spring Harb. Perspect. Biol. 10*, a029181.

Grau-Bové, X., Torruella, G., Donachie, S., Suga, H., Leonard, G., Richards, T.A., and Ruiz-Trillo, I. (2017). Dynamics of genomic innovation in the unicellular ancestry of animals. *Elife 6*, e26036.

Güiza, J., Barría, I., Sáez, J.C., and Vega, J.L. (2018). Innexins: Expression, regulation, and functions. *Front. Physiol. 9*, 1414.

Halfter, W., Oertle, P., Monnier, C.A., Camenzind, L., Reyes-Lua, M., Hu, H., Candiello, J., Labilloy, A., Balasubramani, M., Henrich, P.B., et al. (2015). New concepts in basement membrane biology. *FEBS J. 282*, 4466–4479.

Hall, S., and Ward, R.E. (2016). Septate junction proteins play essential roles in morphogenesis throughout embryonic development in drosophila. *G3 (Bethesda) 6*, 2375–2384.

Hernandez-Nicaise, M.-L., Nicaise, G., and Reese, T.S. (1989). Intercellular junctions in ctenophore integument. In *Evolution of the First Nervous Systems*, P.A.V. Anderson, ed. (Boston, MA: Springer US), pp. 21–32.

Hinman, V., and Cary, G. (2017). The evolution of gene regulation. *Elife 6*, e27291.

Hobert, O., Carrera, I., and Stefanakis, N. (2010). The molecular and gene regulatory signature of a neuron. *Trends Neurosci. 33*, 435–445.

Hoff, M. (2014). Heads, tails, and tools: Morphogenesis of a giant single-celled organism. *PLoS Biol. 12*, e1001862.

Hynes, R.O. (2012). The evolution of metazoan extracellular matrix. *J. Cell Biol. 196*, 671–679.

Jefferson, J.J., Leung, C.L., and Liem, R.K.H. (2004). Plakins: Goliaths that link cell junctions and the cytoskeleton. *Nat. Rev. Mol. Cell Biol. 5*, 542–553.

Johnson, J.L., Najor, N.A., and Green, K.J. (2014). Desmosomes: regulators of cellular signaling and adhesion in epidermal health and disease. *Cold Spring Harb. Perspect. Med. 4*, a015297.

Kania, U., Fendrych, M., and Friml, J. (2014). Polar delivery in plants: Commonalities and differences to animal epithelial cells. *Open Biol. 4*, 140017.

Kapli, P. and Telford, M. J. (2020). Topology-dependent asymmetry in systematic errors affects phylogenetic placement of Ctenophora and Xenacoelomorpha. *Sci. Adv. 6*, eabc5162.

Kenny, N.J., Francis, W.R., Rivera-Vicéns, R.E., Juravel, K., de Mendoza, A., Díez-Vives, C., Lister, R., Bezares-Calderon, L., Grombacher, L., Roller, M., Barlow, L.D., Camilli, S., Ryan, J.F., Wörheide, G., Hill, A.L., Riesgo, A., Leys, S.P. (2020). The genomic basis of animal origins: A chromosomal perspective from the sponge *Ephydatia muelleri*. *bioRxiv.* doi:10.1101/2020.02.18.954784.

Kim, D.H., Xing, T., Yang, Z., Dudek, R., Lu, Q., and Chen, Y.-H. (2017). Epithelial mesenchymal transition in embryonic development, tissue repair and cancer: A comprehensive overview. *J. Clin. Med. 7*.

King, N. (2004). The unicellular ancestry of animal development. *Dev. Cell 7*, 313–325.
King, N., and Rokas, A. (2017). Embracing uncertainty in reconstructing early animal evolution. *Curr. Biol. 27*, R1081–R1088.
King, N., Hittinger, C.T., and Carroll, S.B. (2003). Evolution of key cell signaling and adhesion protein families predates animal origins. *Science 301*, 361–363.
King, N., Westbrook, M.J., Young, S.L., Kuo, A., Abedin, M., Chapman, J., Fairclough, S., Hellsten, U., Isogai, Y., Letunic, I., et al. (2008). The genome of the choanoflagellate Monosiga brevicollis and the origin of metazoans. *Nature 451*, 783–788.
Kular, J.K., Basu, S., and Sharma, R.I. (2014). The extracellular matrix: Structure, composition, age-related differences, tools for analysis and applications for tissue engineering. *J. Tissue Eng. 5*, 2041731414557112.
Le Bivic, A. (2013). Evolution and cell physiology. 4. Why invent yet another protein complex to build junctions in epithelial cells? *Am. J. Physiol. Cell Physiol. 305*, C1193–C1201.
Ledger, P.W. (1975). Septate junctions in the calcareous sponge *Sycon ciliatum*. *Tissue Cell 7*, 13–18.
Lefèbvre, F., Prouzet-Mauléon, V., Hugues, M., Crouzet, M., Vieillemard, A., McCusker, D., Thoraval, D., and Doignon, F. (2012). Secretory pathway-dependent localization of the Saccharomyces cerevisiae Rho GTPase-activating protein Rgd1p at growth sites. *Eukaryot. Cell 11*, 590–600.
Leys, S.P., and Hill, A. (2012). The physiology and molecular biology of sponge tissues. In *Advances in Marine Biology* (Amsterdam: Elsevier), pp. 1–56.
Leys, S.P., and Riesgo, A. (2012). Epithelia, an evolutionary novelty of metazoans. *J. Exp. Zool. (Mol. Dev. Evol.) 318*, 438–447.
Leys, S.P., Cheung, E., and Boury-Esnault, N. (2006). Embryogenesis in the glass sponge Oopsacas minuta: Formation of syncytia by fusion of blastomeres. *Integr. Comp. Biol. 46*, 104–117.
Leys, S.P., Mackie, G.O., and Reiswig, H.M. (2007). The biology of glass sponges. In *Advances in Marine Biology* (Cambridge, MA: Academic Press), pp. 1–145.
Leys, S.P., Nichols, S.A., and Adams, E.D.M. (2009). Epithelia and integration in sponges. *Integr. Comp. Biol. 49*, 167–177.
Lowe, J.S., and Anderson, P.G. (2015). Chapter 3 – Epithelial cells. In *Stevens & Lowe's Human Histology*, 4th Edition, J.S. Lowe and P.G. Anderson, eds. (Philadelphia, PA: Mosby), pp. 37–54.
Magie, C.R., and Martindale, M.Q. (2008). Cell–cell adhesion in the cnidaria: Insights into the evolution of tissue morphogenesis. *Biol. Bull. 214*, 218–232.
Maizel, A. (2018). Plant biology: The making of an epithelium. *Curr. Biol. 28*, R931–R933.
Marioni, J.C., and Arendt, D. (2017). How single-cell genomics is changing evolutionary and developmental biology. *Annu. Rev. Cell Dev. Biol. 33*, 537–553.
Medwig, T.N., and Matus, D.Q. (2017). Breaking down barriers: The evolution of cell invasion. *Curr. Opin. Genet. Dev. 47*, 33–40.
Miller, P.W., Clarke, D.N., Weis, W.I., Lowe, C.J., and Nelson, W.J. (2013). The evolutionary origin of epithelial cell–cell adhesion mechanisms. *Curr. Top. Membr. 72*, 267–311.
Miller, P.W., Pokutta, S., Mitchell, J.M., Chodaparambil, J.V., Clarke, D.N., Nelson, W.J., Weis, W.I., and Nichols, S.A. (2018). Analysis of a vinculin homolog in a sponge (phylum Porifera) reveals that vertebrate-like cell adhesions emerged early in animal evolution. *J. Biol. Chem. 293*, 11674–11686.
Moroz, L.L., and Kohn, A.B. (2016). Independent origins of neurons and synapses: Insights from ctenophores. *Philos. Trans. R. Soc. Lond. B Biol. Sci. 371*, 279–308.
Murray, P.S., and Zaidel-Bar, R. (2014). Pre-metazoan origins and evolution of the cadherin adhesome. *Biol. Open 3*, 1183–1195.
Nagawa, S., Xu, T., and Yang, Z. (2010). RHO GTPase in plants. *Small GTPases 1*, 78–88.

Nichols, S.A., Roberts, B.W., Richter, D.J., Fairclough, S.R., and King, N. (2012). Origin of metazoan cadherin diversity and the antiquity of the classical cadherin/β-catenin complex. *Proc. Natl. Acad. Sci. USA 109*, 13046–13051.

Niklas, K.J. (2014). The evolutionary-developmental origins of multicellularity. *Am. J. Bot. 101*, 6–25.

Olson, B.J. (2013). From brief encounters to lifelong unions. *Elife 2*, e01287.

Özbek, S., Balasubramanian, P.G., Chiquet-Ehrismann, R., Tucker, R.P., and Adams, J.C. (2010). The evolution of extracellular matrix. *MBoC 21*, 4300–4305.

Parfrey, L.W., and Lahr, D.J.G. (2013). Multicellularity arose several times in the evolution of eukaryotes. *Bioessays 35*, 339–347 (Response to doi:10.1002/bies.201100187).

Pozzi, A., Yurchenco, P.D., and Iozzo, R.V. (2017). The nature and biology of basement membranes. *Matrix Biol. 57–58*, 1–11.

Renard, E., Leys, S.P., Wörheide, G., and Borchiellini, C. (2018). Understanding animal evolution: The added value of sponge transcriptomics and genomics. *BioEssays 40*, e1700237.

Reynolds, A.B. (2011). Epithelial organization: New perspective on α-catenin from an ancient source. *Curr. Biol. 21*, R430–R432.

Richter, D.J., and King, N. (2013). The genomic and cellular foundations of animal origins. *Annu. Rev. Genet. 47*, 509–537.

Richter, D.J., Fozouni, P., Eisen, M.B., and King, N. (2018). Gene family innovation, conservation and loss on the animal stem lineage. *Elife 7*, e38726.

Riesgo, A., Farrar, N., Windsor, P.J., Giribet, G., and Leys, S.P. (2014). The analysis of eight transcriptomes from all poriferan classes reveals surprising genetic complexity in sponges. *Mol. Biol. Evol. 31*, 1102–1120.

Rodriguez-Boulan, E., and Nelson, W.J. (1989). Morphogenesis of the polarized epithelial cell phenotype. *Science 245*, 718–725.

Royer, C., and Lu, X. (2011). Epithelial cell polarity: A major gatekeeper against cancer? *Cell Death Differ. 18*, 1470–1477.

Ruthmann, A., Behrendt, G., and Wahl, R. (1986). The ventral epithelium of Trichoplax adhaerens (Placozoa): Cytoskeletal structures, cell contacts and endocytosis. *Zoomorphology 106*, 115–122.

Salinas-Saavedra, M., and Martindale, M.Q. (2019). Par protein localization during the early development of Mnemiopsis leidyi suggests different modes of epithelial organization in Metazoa. *BioRxiv*, 431114.

Satterlie, R.A., and Case, J.F. (1978). Gap junctions suggest epithelial conduction within the comb plates of the ctenophore Pleurobrachia bachei. *Cell Tissue Res. 193*, 87–91.

Schenkelaars Q., Vernale A., Fierro-Constaín L., Borchiellini C., Renard E. (2019). A Look back over 20 years of evo-devo studies on sponges: A challenged view of Urmetazoa. In: Pontarotti P. (eds) *Evolution, Origin of Life, Concepts and Methods*. Springer, Cham.

Schippers, K.J., Nichols, S.A., and Wittkopp, P. (2018). Evidence of signaling and adhesion roles for β-catenin in the sponge *Ephydatia muelleri*. *Mol. Biol. Evol. 35*, 1407–1421.

Sebé-Pedrós, A., and Ruiz-Trillo, I. (2010). Integrin-mediated adhesion complex: Cooption of signaling systems at the dawn of Metazoa. *Commun. Integr. Biol. 3*, 475–477.

Sebé-Pedrós, A., Roger, A.J., Lang, F.B., King, N., and Ruiz-Trillo, I. (2010). Ancient origin of the integrin-mediated adhesion and signaling machinery. *Proc. Natl. Acad. Sci. USA 107*, 10142–10147.

Sebé-Pedrós, A., Degnan, B.M., and Ruiz-Trillo, I. (2017). The origin of Metazoa: A unicellular perspective. *Nat. Rev. Genet. 18*, 498–512.

Sebé-Pedrós, A., Chomsky, E., Pang, K., Lara-Astiaso, D., Gaiti, F., Mukamel, Z., Amit, I., Hejnol, A., Degnan, B.M., and Tanay, A. (2018). Early metazoan cell type diversity and the evolution of multicellular gene regulation. *Nat. Ecol. Evol. 2*, 1176–1188.

Sekiguchi, R., and Yamada, K.M. (2018). Chapter four – Basement membranes in development and disease. In *Current Topics in Developmental Biology*, E.S. Litscher and P.M. Wassarman, eds. (Cambridge, MA: Academic Press), pp. 143–191.

Simion, P., Philippe, H., Baurain, D., Jager, M., Richter, D.J., Di Franco, A., Roure, B., Satoh, N., Quéinnec, É., Ereskovsky, A., et al. (2017). A large and consistent phylogenomic dataset supports sponges as the sister group to all other animals. *Curr. Biol. 27*, 958–967.

Skerrett, I.M., and Williams, J.B. (2017). A structural and functional comparison of gap junction channels composed of connexins and innexins. *Dev. Neurobiol. 77*, 522–547.

Slabodnick, M.M., Ruby, J.G., Dunn, J.G., Feldman, J.L., DeRisi, J.L., and Marshall, W.F. (2014). The kinase regulator mob1 acts as a patterning protein for stentor morphogenesis. *PLoS Biol. 12*, e1001861.

Smith, C.L., and Reese, T.S. (2016). Adherens junctions modulate diffusion between epithelial cells in *Trichoplax adhaerens*. *Biol. Bull. 231*, 216–224.

Smith, C.L., Varoqueaux, F., Kittelmann, M., Azzam, R.N., Cooper, B., Winters, C.A., Eitel, M., Fasshauer, D., and Reese, T.S. (2014). Novel cell types, neurosecretory cells and body plan of the early-diverging metazoan, *Trichoplax adhaerens*. *Curr. Biol. 24*, 1565–1572.

Sogabe, S., Nakanishi, N., and Degnan, B.M. (2016). The ontogeny of choanocyte chambers during metamorphosis in the demosponge *Amphimedon queenslandica*. *Evodevo 7*, 6.

Suga, H., Chen, Z., de Mendoza, A., Sebé-Pedrós, A., Brown, M.W., Kramer, E., Carr, M., Kerner, P., Vervoort, M., Sánchez-Pons, N., et al. (2013). The Capsaspora genome reveals a complex unicellular prehistory of animals. *Nat. Commun. 4*, 2325.

Takaku, Y., Hwang, J.S., Wolf, A., Böttger, A., Shimizu, H., David, C.N., and Gojobori, T. (2014). Innexin gap junctions in nerve cells coordinate spontaneous contractile behavior in Hydra polyps. *Sci. Rep. 4*, 3573.

Tamm, S.L., and Tamm, S. (1987). Massive actin bundle couples macrocilia to muscles in the ctenophore Beroë. *Cell Motil. Cytoskeleton 7*, 116–128.

Tamm, S.L., and Tamm, S. (1991). Reversible epithelial adhesion closes the mouth of beroe, a carnivorous marine jelly. *Biol. Bull. 181*, 463–473.

Tamm, S.L., and Tamm, S. (2002). Novel bridge of axon-like processes of epithelial cells in the aboral sense organ of ctenophores. *J. Morphol. 254*, 99–120.

Tucker, R.P., and Adams, J.C. (2014). Chapter eight – adhesion networks of cnidarians: A postgenomic view. In *International Review of Cell and Molecular Biology*, K.W. Jeon, ed. (Cambridge, MA: Academic Press), pp. 323–377.

Tyler, S. (2003). Epithelium—The primary building block for metazoan complexity 1. *Integr. Comp. Biol. 43*, 55–63.

Walko, G., Castañón, M.J., and Wiche, G. (2015). Molecular architecture and function of the hemidesmosome. *Cell Tissue Res. 360*, 363–378.

Whelan, N.V., Kocot, K.M., Moroz, T.P., Mukherjee, K., Williams, P., Paulay, G., Moroz, L.L., and Halanych, K.M. (2017). Ctenophore relationships and their placement as the sister group to all other animals. *Nat. Ecol. Evol. 1*, 1737–1746.

Williams, F., Tew, H.A., Paul, C.E., and Adams, J.C. (2014). The predicted secretomes of *Monosiga brevicollis* and Capsaspora owczarzaki, close unicellular relatives of metazoans, reveal new insights into the evolution of the metazoan extracellular matrix. *Matrix Biol. 37*, 60–68.

Yathish, R., and Grace, R. (2018). Cell–cell junctions and epithelial differentiation. *J. Morphol. Anat. 2*, 3.

Zihni, C., Mills, C., Matter, K., and Balda, M.S. (2016). Tight junctions: From simple barriers to multifunctional molecular gates. *Nat. Rev. Mol. Cell Biol. 17*, 564–580.

6 Evolution of the Sensory/Neural Cell Types

Sally P. Leys, Jasmine L. Mah, Emma K.J. Esposito

CONTENTS

6.1 Introduction: Historical Perspective 101
6.2 What Is a Neuron? 104
6.3 Characterizing a Neuron in Non-Bilaterians 105
6.4 Sensory Cells and Secretory Epithelial Cells in Ctenophores, Placozoans, and Sponges 109
6.5 Gene Markers of Neurons 115
6.6 Phylogeny, Homology and Bilaterian Bias 117
6.7 Hidden Biology: Sensory/Neural Cells That Are New to Science 119
6.8 Conclusions and Future Directions 119
Acknowledgements 120
References 121

6.1 INTRODUCTION: HISTORICAL PERSPECTIVE

How sensory-neuronal cell types arose has been a major field of study for some time (Anderson, 2015). Ideas on the evolution of the nerve cell, or neuron as it came to be known (Waldeyer, 1891), matured over 150 years following the articulation of the cell theory and Darwin's theory of evolution in the mid-1800s (Anctil, 2015). New technologies and techniques gave new views of neurons, and studies of comparative anatomy and physiology—in particular studies of the nervous and coordination systems in cnidarians, ctenophores and sponges—gave insight into how similar and different neurons were in different lineages.

It took some time for nerves to be accepted as single cells and not syncytia. This was due to the fact that cells overlapped one another and often appeared to observers as anastomosing or fusing where they crossed. As tools advanced, so did the understanding of the cell: new microscopes with lenses of higher magnification and less distortion gave higher resolution images of neuronal cell bodies and their terminal processes. New fixation and staining techniques mastered by earlier workers (e.g. Hertwig and Hertwig, 1878; Ramon y Cajal, 1899) and eventually sectioning and histology showed that the cell processes overlapped but did not fuse with one another. Although it was known that some areas of interaction existed where neurons overlapped, the "synapse" as termed by Sherrington (in Schaefer, 1900) was not actually seen until electron microscope techniques were developed in the mid 1900s (Palade and Palay, 1954; Farquhar and Palade, 1963).

Comparative anatomy and physiological studies of jellyfish and ctenophore sensory and nervous systems have long played a large role in deciphering how neurons might have come about. With time, it became clear to earlier workers that these nervous systems were endpoints of independent experiments in neuronal evolution, not steps in making a vertebrate neuron; though the nervous systems were not obviously centralized like those of vertebrates, they were found to be complex (reviewed in Anctil, 2015). In the latter half of the 20th Century, electrical recordings from hydrozoan, scyphozoan, and even anthozoan cnidarians illustrated that neuronal physiology differs even between taxa in the same class (reviewed in Shelton, 1982; Satterlie, 2019).

In view of these great differences in the morphology and physiology of neurons, there has been, and still is, great interest in how neurons arose. The general idea from earlier studies was that as tissue layers evolved, epithelial cells descended into these new layers, differentiating into muscle, neurons, and inter-neurons. However, there were some differences of opinions about this evolutionary concept. It was debated whether or not multiple different cells independently gave rise to different neurons (Claus, 1878; Chun, 1880) or whether a single epithelial cell type gave rise to different cell types at different times (Kleinenberg, 1872; Parker, 1911; reviewed by Parker, 1919) (Figure 6.1A,B). Another hypothesis comes from Mackie (1970). He discovered that epithelia in many invertebrates are electrically coupled, which led him to propose that the first neurons arose from conducting epithelia. He envisioned that a "starting point" for neurons might have been a contractile epithelial sheet of cells that were coupled metabolically and electrically (Fig. 6.1C). From this tissue, different neurosensory cells and neurons would have evolved, as well as cells specialized for contraction (e.g. muscle). In reviewing the range of electrically coupled tissues found across eukaryotes, he concluded that "conducting ability has been evolved repeatedly and independently in evolution." This view from Mackie (1970) suggests a polyphyletic origin of neural cell types that arose in response to functional needs in different groups. What has been attractive about this scenario (lending to its endurance in the later literature) is that Mackie's hypothesis is about conduction of signals, rather than the morphology of cells. It is multifaceted, with different cells differentiating at different times, in different places, and importantly, he expects that conducting epithelia can evolve multiple times.

The question that is still asked today is whether a common set of tools was repurposed in each lineage or whether different cell types adapted to the function of a neuron independently many times. Knowing more about the cellular components of neurons and other cell types has led to more formalized articulations of these concepts. For example, one idea is that cells have arisen by reuse of modules and that cell lineages might reflect retention of particular types of modules (Achim and Arendt, 2014; Arendt, 2020) or alternatively, that neurons and their specialized properties have been reinvented multiple times, even within a lineage (Moroz, 2009). Today's data could provide an unparalleled toolset to test the hypotheses of neuronal evolution proposed over the last 150 years. One possible hindrance is that

Evolution of the Sensory/Neural Cell Types

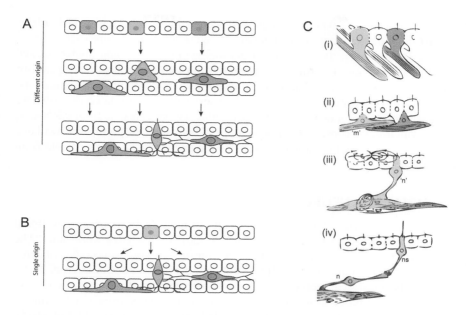

FIGURE 6.1 Hypotheses of neuron origins. (A) Different cell types independently gave rise to muscle (left), sensory neuron (middle), and motor neuron (right), or (B) the same cell type gave rise at different times to each new cell type (after Parker, 1911). (C) A sheet of ciliated epithelial cells which are coupled metabolically and electrically gave rise to effector cells (ii) "protomyocytes" ("m") from different cell types, and of a third type (iii) "protoneurons" ("n") allowing communication between sensor and effector (modified from Mackie, 1970). New effector cell types (neurons, n; neurosensory cells, ns) could arise independently (iv) and connect to muscle cells via chemical synapses. Communication may have been by coupling or by local release of molecules. These illustrations are independent of time or lineage.

the comparative approach on types and components of neurons in different phyla has been lacking in modern work. Although we now have a remarkably in-depth understanding of the molecular biology and physiology of the synapse in mammals and in a handful of invertebrate model organisms, knowledge of molecular biology, physiology of synapses, and the type, location, and content of synaptic vesicles in a range of invertebrate nervous systems, especially non-bilaterians (ctenophores and cnidarians), is lacking by comparison.

Once again, new technologies are helping to alleviate this lack of knowledge. Beginning in 2008, a massive sequencing project's data proposed the possibility that ctenophores, not sponges, are sister to other metazoans (Dunn et al., 2008; Ryan et al., 2013; Whelan, et al. 2015; but see e.g. Feuda et al., 2017; Pett et al., 2019). This hypothesis raised awareness that neurons and nervous systems may have evolved more than once (convergently at least in ctenophores and in the common ancestor of cnidarians and bilaterians) or have been lost entirely more than once in sponges and placozoans (Moroz et al., 2014; Moroz, 2015; Ryan and Chiodin, 2015; Whelan et al., 2015). Another advance is single cell sequencing (scRNAseq), which allows comparison of the transcriptomic signature of cell states and cell types across taxa.

Here, we review the diversity of sensory and neural cells in non-bilaterians and look at transcriptomic signatures of the inferred sensory and neural cell types. Our first point considers what a neuron is and examines the morphology of some of the more characteristic features of neurons and sensory cells in non-bilaterians. Next, we look at gene markers for neurons, and in particular, we evaluate data from scRNAseq studies. And last, we consider the implications of biases in interpreting the available data. These are biases that largely arise from the unequal availability of data from different taxa.

6.2 WHAT IS A NEURON?

There is a universally acknowledged difficulty in identifying a common neuronal phenotype across extant organisms (Anderson, 2015). At a 2014 workshop, "The Evolution of the First Nervous Systems II," a roomful of participants could not agree on a single definition that covered all types of neurons (Bucher and Anderson, 2015). Ion channels and vesicle secretion machinery characteristic of synapses are present in a range of cell types including muscle, osteoclasts, and neurons (Liebeskind et al., 2015, 2016, 2017) as well as in unicellular holozoan ancestors (Burkhardt, 2015); while other cells have the morphological and physiological characteristics of neurons—microtubule-based conduits exist in the excitable syncytia of glass sponges (Mackie and Singla, 1983; Leys, 1995; Leys et al., 1999)—they are not neurons. Some workshop participants proposed that a neuron is a cell that is excitable and can signal chemically, while others suggested a neuron must be polarized, have long extensions (axons and dendrites), and encode information (Rolls and Jegla, 2015; Liebeskind et al., 2016). Using scRNAseq data from a mouse, Huang and Paul (2019) define a neuron as a "neural communication element." The "neural communication element" has two fundamental components, connectivity (partners with which the cell communicates) and the properties of the pre/post synapse, or what they call I/O connectivity. This view defines a neuron by its interaction within a network and by specific aspects of the chemical transmission it carries out, very similar to the morphology-based definitions. A definition that perhaps best combines morphological and physiological attributes is that a neuron is a secretory cell that is polarized and which demonstrates "experience-dependent plasticity and elaborated integrative functions" (Moroz, 2009).

The integrative function of a neuron is what others agree is important, and yet that function relies on a certain, albeit varied, anatomy (Meech, 2017). And so, to discuss the characteristics that are shared by cells that may have given rise to neurons, the neuron is often broken down into morphological components: a receptor, with ion channels or G-protein coupled receptors (GPCR); an excitable membrane with propagating action potential, and hence voltage-gated ion channels; and, most commonly, a chemical synapse with secretory properties. Muscle cells can transmit action potentials (Hernandez-Nicaise et al., 1980), as can epithelia (Mackie, 2004), but they are not considered to be neurons (Anderson, 2015).

Mackie (1970) writes that *"the evolution of conduction is inseparable from the evolution of junctional specialization and... both are subject to the requirements of tissue homeostasis"* (his italics). In recent years, it has become clear that both the transcriptomic signatures and key morphological characteristics of neurons are shaped by the interactions neurons have with other cells. The character of a neuron is affected by the cells it exchanges metabolites and information with, in essence, the neighboring nervous tissues. A similar story applies to other cell types, as is seen in other papers in this volume. Epithelia, excretory cells, or reproductive cell types are each characterized by the composition of the cells that constitute the tissues. Epithelial cells are polarized and have junctions, but the neighbouring cells are not necessarily identical (Renard et al., 2021, this volume). Excretory systems include duct cells and canal cells, and each type rarely occurs alone (Andrikou et al., 2021, this volume). Nervous systems constitute tissues, and the neurons that compose them also rely on other cells, whether they are other neurons, glia, sensory cells, or effector cells (muscle, cilia, or secretory cells). Also, the characteristics of cells in tissues are shared by many other cell types: secretory structures, polarity, cilia, and/or microvilli can all be viewed as modules that are used in many different cell types (Hartwell et al., 1999; Achim and Arendt, 2014). Therefore, what best characterises a cell is the collection of its intrinsic properties as well as the connections it makes with other cells.

6.3 CHARACTERIZING A NEURON IN NON-BILATERIANS

In general, a vertebrate neuron can be characterized as possessing a cell body containing the nucleus, with localized ion channels that shape the outgoing action potential (Duménieu et al., 2017; Leterrier, 2018). The axon internally carries neurotransmitter vesicles to the synapse along microtubule bundles and transmits the action potential along the membrane through the action of voltage-gated ion channels (Rolls and Jegla, 2015). Dendrites contain neurotransmitter receptors as part of the post-synaptic membrane and act to collect and integrate input signals from the pre-synaptic cell (Duménieu et al., 2017). This basic description does not begin to describe the diversity of morphologies, let alone multimodal properties of even GABAergic neurons (Huang and Paul, 2019), but it can serve as a search-pattern to identify a neuron in unfamiliar tissues.

Thus far, the morphology of neurons in invertebrates, with the exception of a few model species, tends to be better characterized than their neuronal transcriptomes (Satterlie and Spencer, 1987; Westfall, 1987; Satterlie, 2019). However, the depth of understanding of morphology varies between organisms. Visualization of tissues can vary greatly, as some organismal tissues can be viewed by light microscope due to the tissues being easily dissected and relatively transparent (Bone 2005). Other tissues can be identified via immunoreactivity to antibodies to neurotransmitters. This technique has been used in jellyfish (Mackie et al., 2003; Satterlie, 2019), acoels (reviewed by Hejnol, 2016), platyhelminths (reviewed by Hartenstein, 2016), and gastropod molluscs (reviewed by Voronezhskaya and Croll, 2016). However, in some tissues, either axons or synapses are the only identifying features, and can only be seen by electron microscopy (Horridge, 1965). But again, new staining, microscopy, and sequencing technologies are beginning to provide a more comprehensive view of the unusual range of sensory neurons and interneurons even in non-bilaterians (Norekian and Moroz, 2019; Norekian and Moroz, 2020).

Synapses are perhaps the cardinal feature of a neuron, and yet they vary considerably in morphology and chemical or electrical character. In what is considered a typical vertebrate synapse, there is a synaptic cleft, or space between two neurons known as the pre- and post-synaptic neurons. The pre-synaptic neuron is the cell that sends signal molecules called neurotransmitters to the post-synaptic neuron, which will then respond based on what receptors are present in the membrane. Different synapses can be characterized by different varieties of adhesion molecules and collections of classical ion channels and vesicle transport proteins. In general, the processes that occur at a synapse involve the packaging of the neurotransmitter into vesicles in the pre-synaptic neuron, release of that neurotransmitter into the synaptic cleft, and receiving of the neurotransmitter by the receptors on the post-synaptic membrane. To compare how diverse synapses can be, we can look at a glutamatergic and a GABAergic synapse. The glutamatergic synapse involves the release of the neurotransmitter glutamate, the major excitatory neurotransmitter in vertebrates. The packaging of this neurotransmitter into vesicles is dependent on the vesicular glutamate transport proteins (vGLUTs) (Wojcik et al., 2004). In contrast, a GABAergic synapse involves the inhibitory neurotransmitter, GABA (γ aminobutyric acid), being packaged into vesicles by vesicular GABA transport proteins (VGATs) (McIntire et al., 1997). Once either neurotransmitter is released into the synaptic cleft, it diffuses towards the post-synaptic neuron. The post-synaptic neuron membrane contains receptor proteins that respond to specific neurotransmitters. There are two general types of receptors, metabotropic and ionotropic, that can be further characterized into subtypes (Niciu et al., 2012). Typically, metabotropic receptors are coupled with secondary signaling proteins, such as G-proteins, that modulate cellular activity via intracellular signaling partners (Niswender and Conn, 2010). This pathway usually results in a slower cellular response. In contrast, the ionotropic receptors, once bound to a specific ligand, undergo conformational changes and allow ions to immediately enter or exit the cell. Both types of receptors have been well-characterized in vertebrates but not with the same intensity in invertebrates, and very little is known about them in non-bilaterians.

Neurons can also be identified by the particular genes they express; however, some genes that are usually associated with synapses are also expressed in other cells that have adhesive and secretory functions. Some of these genes have been identified in sponges and placozoans and are often highlighted as indicating a potential "starting point" for the synapse. The gene involved in the pre-synaptic and post-synaptic density (PSD) of a "standard" glutamatergic vertebrate neuron are identified as present or absent in sponges and placozoans (Figure 6.2A). The scaffolding at the post-synaptic (receiving) side typically contains adhesion molecules including cadherin, alpha, and beta catenin, which link with Homer and Shank and interact via IP_3 signalling. The glutamatergic neuron would have iGluRs as well as voltage gated potassium channels (as well as other K channels), and the GABAergic synapse would have GABA receptors. Pre-synaptic proteins for vesicle transport and

Evolution of the Sensory/Neural Cell Types

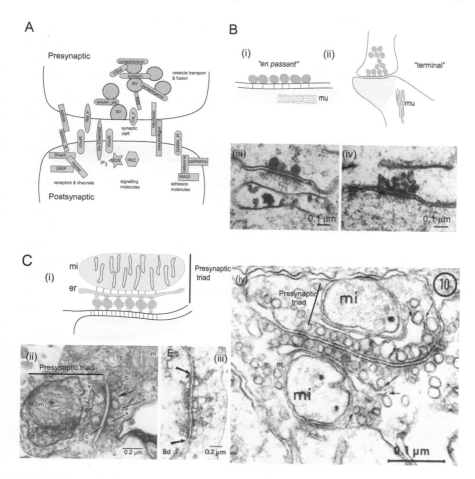

FIGURE 6.2 Diversity of chemical synapse architecture. (A) Layout of a "standard" glutamatergic vertebrate neuronal synapse showing the basic complement of proteins involved in scaffolding, ion signalling, and vesicle transport with those proteins known to be present in sponges shown in green. (B) Sketch (i, ii) and electron micrographs (iii, iv) of *en passant* and terminal type synapses. Electron micrographs show neuro-neuronal (left) and neuro-muscular (right) synapses in *Polyorchis penicillatus*. Reprinted with permission from Spencer and Schwab, Electrical Conduction and Behaviour in 'Simple' Invertebrates, (Oxford, Clarenden Press, 1982). Hydrozoan synapses lack any membrane specializations (folds) and are *en passant* form, whereas anthozoans have *en passant* and terminal synapses. (C) Sketch (i) of a pre-synaptic triad type of synapse in which the endoplasmic reticulum lies between the synaptic vesicles and a mitochondrion. Electron micrographs of synapses in (ii, iii) scyphozoan (*Cyanea capillata*) and (iv) ctenophore (*Beroe ovata*). (m, mitochondria, e, elongate cisternae, b, bulbous cisternae of the endoplasmic reticulum, v, synaptic vesicles, c, synaptic cleft). Scyphozoans have both pre-synaptic triad type (ii) and non-polarized (iii) synapses. Reprinted with permission from Anderson and Grünert, Synapse 2: 603–613 (Wiley, 1988) and Anderson and Schwab, Journal of Morphology 170: 383–399 (Wiley, 1981). In ctenophores, the pre-synaptic triad is typical (iv). Reprinted with permission from Hernandez-Nicaise, Journal of Neurocytology 2(3): 249–263. (Springer, 1973).

docking include RIM, SNAP/SNAREs, synaptobrevin, syntaxin, synaptogamin, and solute carriers as indicated above. Many, but not all, of these families are found in both placozoan and sponge genomes and may be used in a range of tasks that include secretion, uptake, and adhesion. However, the function of signaling to other cells should not be ruled out.

In cnidarians, synapses are more variable and tend to be described by their connectivity to other cells and tissues (neuroepithelial, neuromuscular, neuro-cnidocyte, and neuronal) (Westfall, 1987). Electrical synapses (gap junctions) are present in hydrozoans (Spencer and Schwab, 1982), and although dye coupling between cells was demonstrated in the anthozoan *Renilla koelleri* (Germain and Anctil, 1996), there is so far neither ultrastructural, dye coupling, nor molecular evidence for gap junctions in the other two cnidarian classes (Scyphozoa and Cubozoa). Chemical synapses vary considerably in morphology across cnidarians (Figure 6.2B,C) (Shelton, 1982; Westfall, 1987). In hydrozoans, all synapses are *"en passant"* (Figure 6.2B), meaning that the synapse does not reflect a specialized region of the cell membrane such as a synaptic fold or bouton, but occurs where neurons cross one another (Spencer and Schwab, 1982). The synaptic cleft is 12–20 nm with 80–120 nm diameter horseshoe-shaped vesicles with lightly staining material at neuro-muscular synapses, and 60–150 nm diameter darkly staining vesicles at neuro-neuronal synapses (Spencer and Schwab, 1982). Hydrozoans have asymmetrical (polarized) synapses characterized by a post-synaptic density on the receiving cell membrane associated with scaffolding proteins (Westfall, 1987; Torrealba and Carrasco, 2004; Boeckers, 2006) as well as bidirectional (reciprocal or symmetrical) synapses, in which one end of the synapse is pre-synaptic and the other end is post-synaptic (Spencer and Schwab, 1982). Anthozoans have both *en passant* and terminal synapses (onto muscle cells) (Westfall, 1987) (Figure 6.2B).

Many studies confirm that cnidarian neurons label with antibodies to a range of neuropeptides but not to monoamines or acetylcholine (Grimmelikhuijzen and Westfall, 1995; Satterlie, 2019). Studies on *Hydra* have illustrated that peptide-gated ion channels (*HyNaC*) are almost exclusively used for fast neuromuscular transmission, and their presence in other cnidarians and in ctenophores and placozoans is taken to suggest this might be the case also in those groups (Assmann et al., 2014). Although no previously known neuropeptide classes were found in the *Pleurobrachia* ctenophore genome (Moroz et al., 2014), novel peptide prohormone genes are present and are expressed in the mouth, tentacles and polar fields suggesting a sensory function (Moroz et al., 2014). Ctenophores also have genes encoding for degenerin (Deg)/ENaC peptide-gated channels (Gründer and Assmann, 2015; Moroz and Kohn, 2015), and therefore, while ctenophore synapses are considered to use a diversity of glutamate channels as well as gap junctions for signalling, it is likely that peptide-gated ion channels are also used (Moroz and Kohn, 2015).

Synapses in ctenophores are distinct in ultrastructure from those of hydrozoans but similar to those in scyphozoan jellyfish. The structure is called a "pre-synaptic triad" due to the endoplasmic reticulum (ER) lying between the synaptic vesicles and a mitochondrion (Hernandez-Nicaise, 1973) (Figure 6.2C). The vesicles are attached to the cell membrane and to the ER sac by thin projections. The synaptic cleft is 12–15 nm wide at interneural synapses and 15–20 nm wide at neuro-muscular synapses (Westfall, 1987). In scyphozoans, the pre-synaptic triad is clearly present but does not seem to be standard, as some synapses lack the three parts (e.g. Figure 6.2C) (Anderson and Schwab, 1981; Anderson and Grünert, 1988). In ctenophores, the pre-synaptic triad is standard and some are polarized, others non-polarized, and yet others are bi-polarized (reciprocal) (Hernandez-Nicaise, 1973); however, ctenophore synapses are unusual in that any part of the membrane can form a synapse (Tamm, 1982).

Neither sponges nor placozoans have neurons, although over the years, several candidate cells have been proposed as neurons in sponges (Pavans de Ceccatty, 1955; Pavans de Ceccatty et al., 1970) and arguments against the histological evidence have been discussed at length (Jones, 1962; Pavans de Ceccatty, 1974, 1979). Sponge cells can have juxtaposed membranes with vesicles nearby (reviewed in Leys, 2007), but so far there is no conclusive evidence that these are involved in the kind of rapid information transfer that is characteristic of a synapse. Also, no specific cells have been found to label for peptide, monoamine, or acetylcholine neurotransmitters. Placozoans, in contrast, do label for peptides, and there is new and growing data to suggest that *Trichoplax* releases and responds to the release of very specific secreted peptides (see next section).

6.4 SENSORY CELLS AND SECRETORY EPITHELIAL CELLS IN CTENOPHORES, PLACOZOANS, AND SPONGES

Ctenophores have a number of different types of sensory receptors, all of which are connected directly to neuronal networks (Figures 6.3 and 6.4). The most well-known sensory structure in ctenophores is the apical organ (Satterlie and Spencer, 1987) (Figure 6.3A–-D), which is essentially a gravity receptor, in which calcium granules produced by a lithocyte are held on top of groups of 50 μm long cilia (Tamm, 1982). The balancer cells synapse with neurons at their base (Hernandez-Nicaise, 1973), and the mechanism by which changes in orientation of the balancer organ is transmitted into changes in beat pattern of the comb row cilia have been well-described by Tamm (1982). Other sensory receptors vary in type, location, and structure, even between closely related species of ctenophore (Hernandez-Nicaise, 1974; Norekian and Moroz, 2019) (Figure 6.4). Long non-motile stiff cilia with an onion-root like base (Horridge, 1965) are found on the tips of finger-like projections (Tastborsten of Hertwig, 1880) in some ctenophores (Figure 6.4A–C). The most common type of receptor, however, seems to be a sensory peg, a stiff projection which is formed by a dense set of actin microfilaments, which are thought to be receptors of water movement (Horridge, 1965) (Figure 6.4D,E). Similar fibrillar pegs and solitary cilia are seen on the surface of *Euplokamis dunlapae,* and slightly different fibrillar sensory receptors lie among colloblasts on the tentilla (Norekian and Moroz, 2020). In *Beroe abyssicola*, there are at least five potential sensory receptor cell types, several of which stain strongly for actin like the fibrillar pegs (Norekian and Moroz, 2019)

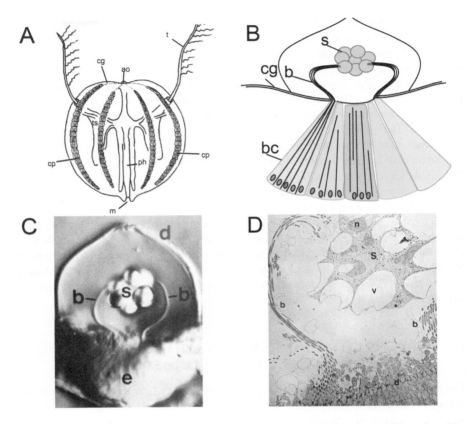

FIGURE 6.3 Sensory cells in ctenophores: the apical organ. (A) Diagram of *Pleurobrachia bacheii*. Apical organ (ao), ciliary groove (cg), comb plate (cp), tentacle sheath (ts), pharynx (ph). Sketch (B), light micrograph (C), and electron micrograph (D), of balancer cells (bc/e) and balancer cilia (b) supporting the statocysts (s). Changes in position are translated to the comb row cilia via the ciliated grove (cg). A,C,D Reprinted with permission from Tamm, Electrical Conduction and Behaviour in 'Simple' Invertebrates, (Oxford, Clarenden Press, 1982).

(Figsure 6.4F,G). In all species of ctenophore studied so far, clusters of ciliated cells forming "rosettes" occur in the meridional canals that underlie the eight comb rows.

Because it is clear which cells give rise to the apical organ in ctenophores, it has also been possible to identify which genes are involved in forming that structure and the neurons associated with it. In *Mnemiopsis leidyi*, the LIM homeobox genes *MlLhx3/4* and *MlLhx4/5* are expressed in cells from which the apical organ derives and the presumptive photocytes respectively (Simmons et al., 2012). In *Pleurobrachia pileus*, SOX E genes are expressed in the cells that form the floor of the apical organ, while SOX B group genes are expressed in cells that give rise to the polar fields around the apical organ (Jager et al., 2011). However, in *M. leidyi*, expression of Sox genes in those same locations was interpreted to reflect a role in

Evolution of the Sensory/Neural Cell Types 111

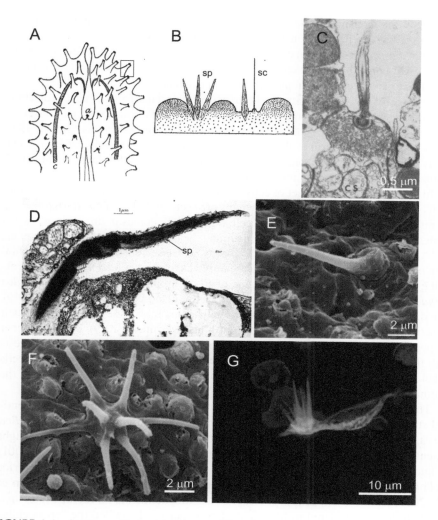

FIGURE 6.4 Sensory cells in ctenophores: cilia and sensory pegs. (A-B) Diagrams of the projections ("fingers") on the surface of *Leucotheca multicornis* and sensory pegs (sp) and sensory cilia (sc) on the surface of the fingers. (C) Sensory cilium, seen by transmission microscopy. Reprinted with permission from Hernandez-Nicaise, Tissue and Cell 6(1): 43–47 (Wiley, 1974). (D-G) Different views illustrating the actin-dense nature of the sensory peg: (D) Transmission electron microscopy, (E-F) scanning electron microscopy, and (G) fluorescence microscopy (red, actin labelled by phalloidin; green, tubulin, labelled with acetylated alpha tubulin; blue, nuclei, labelled with Hoechst). A,B,D Reprinted with permission from Horridge 1965, Proceedings of the Royal Society of London, B, 162(988): 333–350. E-G, Reprinted with permission from Norekian and Moroz, Journal of Comparative Neurology 527(12): 1986–2008 (Wiley, 2019).

stem cell maintenance (Schnitzler et al., 2014). The contrasting interpretations of gene expression reflects a general hesitation to use bilaterian molecular markers to identify sensory structures in ctenophores. As Simmons et al. (2012) note, the ctenophore LIM homeobox complement is more similar to sponges, which lack neurons, than it is to cnidarian or even placozoan complements.

A range of complex sensory cells exist in cnidarians from ciliary projections of cnidocytes and epitheliomuscular cells (Golz and Thurm, 1994; Galliot et al., 2009) to statocysts, eye spots, and the complex eyes of cubozoans (Singla, 1974; Nilsson et al., 2005; Garm et al., 2008). These have been extensively reviewed previously (Passano, 1982; Shelton, 1982; Spencer and Schwab, 1982; Jacobs et al., 2007; Satterlie, 2019) and are therefore not discussed here.

In placozoans, a diverse set of cells make up the upper and lower layers. A "lipophil" cell has large vesicles whose contents are secreted to lyse algae during feeding (Smith et al., 2015). Another type of cell, a gland cell, comes in three forms, two of which (types I and II) are concentrated around the border of the animal (Smith et al., 2014, 2015; Mayorova et al., 2019) (Figure 6.5). These cells release endomorphin-like peptides which appear to control the animal's feeding behavior (Senatore et al., 2017; Mayorova et al., 2019). Expression of *ELAV* around the margin of *Trichoplax adhaerens* correlates with the distribution of the gland cells (DuBuc et al., 2019). We do not know if the upper ciliated epithelial cells may be sensory, or what their function might be (Smith et al., 2015; Smith and Mayorova, 2019), but the number of peptides that placozoans respond to, each producing distinct behaviors, suggests that the ventral epithelial cells have multiple receptors. In some cases, distinct cells label for distinct peptides, producing a mosaic of receptors across the surface of the animal that is in contact with the substrate (Varoqueaux et al., 2018).

The sponge larva also has sensory-effector cells that respond to light and chemical stimuli. At the posterior pole of the demosponge *Amphimedon queenslandica* larva, a ring of cells with long cilia bend or extend in response to changes in light intensity, steering the larva in a particular direction (Leys and Degnan, 2001) (Figure 6.6 A-D); other demosponge larvae also have long cilia at the posterior pole which respond in a similar way to changes in light intensity (Maldonado et al., 2003; Collin et al., 2010). The calcareous amphiblastula larvae of *Sycon spp.* has long been considered to have four photosensory cells called the "cross cells" (Bidder, 1937; Tuzet, 1970; Lanna and Klautau, 2012), but how they function as effectors is still unknown. The *Pax-Six-Eya-Dach* network (PSEDN) and *Sox* family genes have been found in the calcarean sponges *Sycon ciliatum, Sycon calcavaris*, and *Leucosolenia complicata* (Fortunato et al., 2012, 2014) and in demosponges *Chalinula loosanoffi* and *Ephydatia syriaca* (previously *E. fluviatilis*) as well as *A. queenslandica* (Hoshiyama et al., 1998; Hoshiyama et al., 2007; Larroux et al., 2008; Hill et al., 2010). The cross cells of *Sycon* express *Pax, Six*, and *Sox* and also *ELAV* and *Msi* (Fortunato et al., 2014). Some of these transcription factors, like *Sox*, may be markers of the stem cell state of the cross cells at that point in the development of the larva, but expression of *ELAV* and *Msi* in the cross cells and posterior pole macromeres is interpreted as being associated with their putative sensory function (Fortunato et al., 2014).

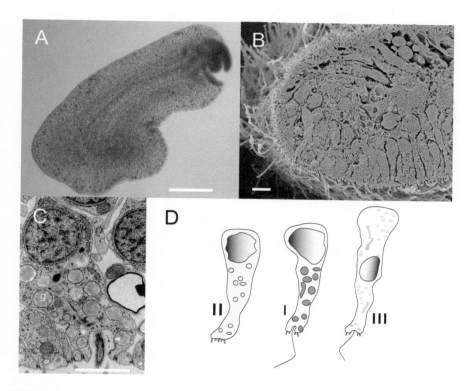

FIGURE 6.5 Sensory cell types in the placozoan *Trichoplax adhaerens*. (A) Light micrograph of the whole animal (Leys, Friday Harbor Embryology course, 2006). The gland cell in *Trichoplax adhaerens* is characteristic of the ventral-lateral edge and has characteristic inclusions that label for peptide signalling molecules. (B) Fracture of *Trichoplax* viewed by scanning electron microscopy showing two types of gland cells; G_I, G_{II} are highlighted in blue and pink (Leys, unpublished). (C) Transmission electron micrograph of one of the gland cells. Reprinted with permission from Smith et al. Current Biology 24(14):1565–1572 (Elsevier, 2014). (D) Cartoon, showing the three types of gland cells identified by Mayorova et al. (2019). Scales A, 100 µm, B, 2 µm, C, 2 µm.

In the *A. queenslandica* sponge parenchymella larva, a flask-shaped cell occurs among the columnar ciliated cells and is more common in the anterior half of the larva (Leys and Degnan, 2001; Nakanishi et al., 2015) (Figure 6.6 E-G). It is thought that the flask cell receives chemical signals from algae and sends signals via calcium and nitric oxide to adjacent cells, triggering metamorphosis (Nakanishi et al., 2015; Ueda et al., 2016). A cell adjacent to the flask cell, called the "globular cell," expresses the transcription factors *bHLH* and *Notch/Delta* during development (Richards et al., 2008; Richards and Degnan, 2012). Expression of *AmqbHLH1* in *Xenopus* induces ectopic expression of n-tubulin, a marker of neural differentiation, and in *Drosophila,* expression of *AmqbHLH1* induces the expression of ectopic sensory bristles on the wing (Richards et al., 2008). These heterologous expression

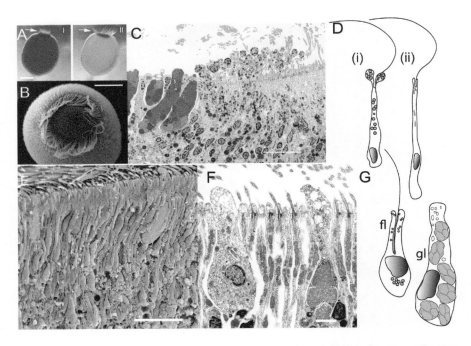

FIGURE 6.6 Several sensory cell types in the demosponge *Amphimedon queenslandica*. (A-D) Light responsive cilia at the posterior pole of *A. queenslandica*; (E-G) flask and globular cells at the surface of the *A. queenslandica* larva. (A) Positions of the posterior cilia when light is turned down (I) and up (II) (from Leys et al. 2002). (B) Scanning electron micrograph of the posterior pole of the *A. queenslandica* larva showing cilia folded down over the pole (Leys, unpublished). (C) Cross section, TEM, of the posterior pole of the *A. queenslandica* larva showing large cells with gray inclusions on the left that form the posterior pole, and to the right, ciliated cells which contain pigment granules in bulbous extensions (Leys, unpublished). (D) Diagrams of the cells with pigment inclusions and neighboring cells with long cilia that lack the pigment inclusions. (E) Flask cell and globular cell, highlighted in pink and blue, respectively, on the surface of the larva. (F) Transmission electron micrograph of the flask (left) and globular (right) cells (Leys, unpublished). (G) Cartoon illustrating the characteristic shape of the flask cell (fl) with deeply recessed cilium, clear vesicles in the apical side, and darker vesicles in the basal side of the cell and globular cell (gl) with large gray inclusions. Scales A,B 100 μm; C,E 10 μm; F, 2 μm.

experiments indicate that the sponge transcription factor is able to carry out the role of determining sensory structures in bilaterians.

Another sensory cell in sponges is the hair cell, which appears sporadically in the sponge excurrent canal system and lines the inside of the whole osculum (Boury-Esnault and Rützler, 1997; Elliott, 2009; Nickel, 2010; Ludeman et al., 2014). There is evidence that these cells function to sense movement of water through the osculum. Removal of the cilia or of the osculum or blocking calcium channels with drugs that block hair cell calcium channels, all prevent typical behaviors by the sponge (Ludeman et al., 2014). Endopinacocytes, the cell type that lines the canals

Evolution of the Sensory/Neural Cell Types   115

and which can be ciliated, also express *Pax 2/5/8* (Hill et al., 2010; Rivera et al., 2013; Hall et al., 2019).

While both cnidarian and ctenophore receptors are clearly associated with neurons that are connected to a network and which transfer information, this is not the case in either sponges or placozoans, although there is good evidence for paracrine signalling between cells to coordinate behavior in both of those groups.

6.5 GENE MARKERS OF NEURONS

Neurotransmitters have long been used to identify neurons in a range of tissues, from vertebrates to jellyfish (Torrealba and Carrasco, 2004). They are useful markers, but not all neurons label even in a single animal, and although developing custom antibodies has been a typical approach, these still do not identify all neurons (Grimmelikhuijzen and Westfall, 1995), and in some cases, the antibodies can also label non-neuronal cells (e.g. Faltine-Gonzalez and Layden, 2019). Transcription factors are typically another useful marker, and many label nascent neurons in tissues (Marlow et al., 2009; Jager et al., 2011; Layden et al., 2012; Nakanishi et al., 2012); however again, different transcription factors are expressed in different species with only a few transcription factor types being useful cross-species and cross-phyla markers even in cnidarians (reviewed in Galliot et al., 2009). As noted previously (Tamm 1982), classical markers for neurons are "unreliable" in ctenophores, and this seems also true for transcription factors. In *Pleurobrachia bachei*, no prohormone-derived peptides were found with homology to any cnidarian+trichoplax+bilaterian peptide, but 72 novel prohormone peptides were found which, as described earlier (Section 3), were expressed around the mouth and tentacles and in the polar fields (Moroz et al., 2014). While transcription factors involved in neurogenesis in cnidarians and bilaterians (e.g. *bHLH*, *Sox*, *Six*, and *Gli*) as well as RNA binding molecules involved in stem cell and neural differentiation (*ELAV*, *Msi*) are present (Ryan et al. 2013), *in situ* hybridization experiments show that some, like *ELAV*, are expressed in muscle cells, and not in neurons (Moroz et al.s 2014). Grasping at individual markers that are not necessarily reliable, even across a phylum, makes identifying the origins of neurons in these non-bilaterian groups challenging. However, the development of new techniques promises new steps forward, and single cell RNAseq methods offer one new approach.

The scRNAseq approach with invertebrate model species, such as *Drosophila* (Croset et al., 2018) and *C. elegans* (Cao et al., 2017; Lorenzo et al., 2019), is providing massive, rich datasets for future comparative analyses. In *Drosophila*, neuronal cell populations can be isolated so that the identity of cell clusters as neurons is known beforehand. In *C. elegans*, there are significantly fewer neurons, so the approach is to first identify these from the remaining cells, using gene markers, and then sequence them.

A gene "marker" refers to genes that are enriched in a cell cluster (i.e. are used to label or identify cell clusters). They can include transcription factors but often consist of protein coding genes that can be involved in vesicle transport, docking, or vesicle content (e.g. neurotransmitter types). Transcription factors are more likely to identify a cell cluster in differentiating tissues, a so-called "transcription résumé" (Paul et al., 2017). Neuronal cell populations do not always cluster by the same markers in different organisms, and often clusters are identified instead by using previously known neuronal cell markers. For example, in *Drosophila*, cells in the mushroom bodies are identified by their expression of the Kenyon Cell markers *eyeless* and *Dopamine receptor in mushroom bodies* (*damb, or Dop1R2*) (Croset et al., 2018). In *C. elegans*, neuronal cells were identified in clusters by their expression of the neuronal markers *egl-21*, *unc-104*, and *unc-10* (Cao et al., 2017). Although neuronal cells express distinct sets of genes in different species, the functional diversity of populations of cells can still be seen in the sets of genes expressed within a species.

In light of such diversity of expression and of the distinct pan-neuronal markers used for each of these species, it is not clear that scRNAseq data are going to be an effective tool for honing in on a universal neuronal marker (or set of markers) for bilaterians, let alone for non-bilaterians. In the anthozoan cnidarian model, *Nematostella vectensis* neurons are dispersed throughout the body so that scRNAseq must be carried out on the whole animal to differentiate cell types, including neurons (Sebé-Pedrós et al., 2018b). Sebé-Pedrós et al. (2018b) sorted 11,888 cells from *N. vectensis*. Since the cells were of different sizes, variance in gene expression was used to identify 700 marker genes with 104 transcriptional signatures (Sebé-Pedrós et al., 2018b). These markers grouped cells into "metacells," which are defined as groups of "scRNA-seq cell profiles that are statistically equivalent to samples derived from the same RNA pool" (Baran et al., 2019). In *N. vectensis*, neuronal metacell clusters were identified by the presence of transcription factors, specifically *ELAV* and *ASCc* (*achaete-scute* homolog), and non transcription factor marker genes. The authors identified 32 neuronal metacells with hundreds of markers, indicating functional diversity of the neuronal cell populations. They concluded that neuronal cell populations were rich in genes that had a range of origin times, which supports the idea that the neuron is composed of modules (Achim and Arendt, 2014) that became assembled at different timepoints in animal evolution. They also concluded that while neuronal cell populations in *N. vectensis* have broad similarities (e.g., expression of *ELAV* and *ASCc*), the diversification of markers expressed in these populations (e.g., *Fox2*, *Pou*, *Shaker*, *Otp*, and RFamide in some, and *Gata*, *Otx*, *Islet*, LWamide in others) suggests that "neural radiations" assembled specific (and distinct) gene modules even within the cnidarian lineage (Sebé-Pedrós et al., 2018b).

It is interesting that the *N. vectensis* scRNAseq study was able to confirm one of the two fundamental features of neurons defined by Huang and Paul (2019), namely the chemical signature of the synapse. Of the 32 neuronal metacell populations identified, some were characterised by differences in peptides (RFamide or LWamide), and these peptides did localize to distinct tissues in the polyp. It seems that the presence of characteristic neurotransmitters is still one of the most appropriate defining features of a neuron (Sebé-Pedrós et al., 2018b).

Single cell RNA sequencing data from non-bilaterians that lack neurons, e.g. sponges and placozoans, and in ctenophores which have unconventional nervous systems, found no metacells with a distinct neuronal expression signature

Evolution of the Sensory/Neural Cell Types

(Sebé-Pedrós et al., 2018a). Even in the ctenophore *Mnemiopsis leidyi*, Sebé-Pedrós et al. (2018a) found that components involved in a classical vertebrate synaptic scaffold (as described earlier in Section 3) were expressed in many different cell types, and no single cluster co-expressed many voltage gated ion channels (Sebé-Pedrós et al., 2018a). In *M. leidyi*, many metacells expressed innexins (components of gap junctions), as well as ASC, iGlur, and the voltage gated sodium, potassium, and calcium channels. Moreover, in the placozoan *Trichoplax adhaerens*, which has no nervous system but which does have cells that secrete and respond to peptides, genes considered to have a neuronal role in bilaterians were not co-expressed with peptidergic cell clusters. The authors concluded that this indicated "the absence of a synaptic scaffold or any other neuronal gene module" in placozoans (Sebé-Pedrós et al., 2018a). In the sponge, placozoan, and ctenophore models studied, *ELAV* was not associated with any signalling component, unlike in cnidarians (Sebé-Pedrós et al., 2018a), and not a single transcription factor indicated a neuronal cell cluster. Conclusions from scRNAseq analysis were that a) in sponges and placozoans, no collection of genes particularly marked a neuronal cell type precursor, and b) markers for neuronal signalling in ctenophores are sufficiently different from cnidarians and bilaterians to lend support to hypotheses that neurons originated independently at different times (Moroz, 2009; Moroz et al., 2014; Moroz, 2015). In summary, there is no neuronal transcriptional signature in early branching non-bilaterians (sponges, placozoans, and ctenophores), but in cnidarians, *ELAV* is considered one of the few signatures of neuronal cell populations, and transmitter molecule types characterize populations of neurons in cnidarians as they do in bilaterians.

6.6 PHYLOGENY, HOMOLOGY AND BILATERIAN BIAS

It is often written that the position of ctenophores as sister to remaining metazoans indicates that sponges lost neurons or that there were multiple origins of neurons (Ryan and Chiodin, 2015; Whelan et al., 2015; Moroz and Kohn, 2016; Telford et al., 2016). But because the data suggest that transmitter types and conducting ability have evolved repeatedly in different lineages, the position of sponges or ctenophores in metazoan phylogeny is not informative about the deep roots of nervous system development. First, there are no pan-neuronal transcription factors. The expression of transcription factors such as *ASCc*, *ELAV*, and *Pax* often patterns developing neuronal tissues in cnidarians and bilaterians, but as these transcription factors also have other roles and do not label neurons in ctenophores (Sebé-Pedrós et al., 2018a), they are not unambiguous, pan-neuronal markers. Second, neurons can arise from ectodermal, endodermal, or mesodermal tissues in different phyla (Hartenstein and Stollewerk, 2015) and so do not share an embryological origin across phyla. Third, while the identifying character of vertebrate neurons is their transmitter complement (Huang and Paul, 2019), and peptide neurotransmitters can also be identified in neurons of the cnidarian *Nematostella* (Sebé-Pedrós et al., 2018b), many analyses show that the metabolism of transmitters diversified independently in different lineages that had already diverged hundreds of millions of years previously (Liebeskind et al., 2016; Moroz and Kohn, 2016; Liebeskind et al., 2017; Faltine-Gonzalez and

Layden, 2019). Similarly, ion channel complement has been found to be lineage specific, reflecting diversification (or loss) in phyla (Liebeskind et al., 2015; Fernández and Gabaldón, 2020). While there is clearly no pan-neuronal marker, deep homology (Shubin et al., 2009) of some of those transcription factors cannot be ruled out; hence the interest in the frequent association of *bHLH* or *ELAV* with sensory cells in aneural non-bilaterians (Richards et al., 2008; Fortunato et al., 2014, 2016; DuBuc et al., 2019).

Can a single cell sequencing approach eventually work to confidently identify a neuronal cell type? The advantage of using single cell approaches is that the genes that statistically cluster cells (i.e. assign each cell to a cluster) are chosen empirically, and these are the genes whose expression levels fluctuate the most from cell to cell (Achim et al., 2015). Genes that have low variance, such as housekeeping genes, would be equally expressed among all clusters and do not define the boundaries of clusters. Ideally the gene set defining the cluster boundaries would be transcription factors, such as those indicated previously. In reality, however, a number of different marker genes define the clusters (Sebé-Pedrós et al., 2018a), and many of these are effectors, such as receptors or neurotransmitters. Genes involved in the construction and functioning of the synapse are also effectors; therefore, it is unsurprising that they show up in all metacell clusters of ctenophore, placozoan, and sponge datasets (Sebé-Pedrós et al., 2018a).

It must also be admitted that the method of defining cell clusters is slightly circular. Either genes already known to be expressed in the cells are used to label clusters (such as the example of *unc-10* and other neuronal genes identifying clusters of *C. elegans* cells) (Lorenzo et al., 2019), or morphological characters of the cell types are used to choose marker genes; for example, genes known to be involved in the flagella apparatus and in microvilli are used to identify choanocytes (Sebé-Pedrós et al., 2018a). But without the morphological correlate, novel cell types are difficult to identify. In cases where a cluster of cells is seen, and the marker genes that identify it do not match a known phenotype, the cluster is ascribed to a known morphological or chemical phenotype regardless (e.g. two types of choanocytes) (Sebé-Pedrós et al., 2018a; Zimmermann et al., 2019). The resolution of scRNAseq methods may be impressive where neurotransmitter identity is extremely well known, as in the case of mouse neocortical neurons (Huang and Paul, 2019), but where the chemical, physiological and morphological knowledge space is blank, as it is for many early branching metazoans, scRNAseq cell identities are poorly resolved compared to what is possible by ultrastructure. Some 31 cell types have been identified morphologically in some sponges (Simpson, 1984, pp. 92–98, 100; Tables 3–4, 3–5), whereas only 10 sorted out by scRNAseq (Sebé-Pedrós et al., 2018a).

Finally, the often unspoken problem in identifying a pre-neuronal cell type in any animal is the bilaterian bias (Dunn et al., 2015). Marker genes are known from synapses in bilaterians. The more distantly related the animal, the less likely it is to possess a full complement of marker genes, giving the impression that complexity increases from non-bilaterians to bilaterians. Looking for markers of synapses with architecture such as the "pre-synaptic triad" found in scyphozoans and ctenophores might give a different picture, but how would a completely unconventional synapse be identified? At what point can the word synapse still be applied to the juxtaposition of two membranes and the secretion of substances from vesicles? If the widely used metabolite glutamate or NO (nitric oxide) and ATP were the transmitters of "protoneurons" (Moroz, 2009), it would be difficult to find cells specializing in using them only for signalling to other cells and for specific behaviors.

Anderson (2015) writes that "we know a neuron when we see one," but if we have a bilaterian or parahoxozoan (Ryan et al., 2010) bias by looking for neurotransmitters, polarity, or elongate processes, then how would we identify a totally novel cell type that received and transmitted information to another cell? It is important to be conscious of bilaterian bias in thinking about the ancestry of the sensory-neural cell type.

6.7 HIDDEN BIOLOGY: SENSORY/NEURAL CELLS THAT ARE NEW TO SCIENCE

If the flask cell in *A. queenslandica* is a neuronal analog, why would other sponges not have the same cell type and signalling mechanism? Many epithelial cells could have converged on the same sensory cell morphology, and the expression of transcription factors in these cells may also be convergent. The ctenophore *Leucotheca multicornis* has a very similar set of sensory pegs and sensory cilia on finger-like processes arising from the body to those of *Beroe abyssicola*, despite the two species belonging to separate orders (Horridge, 1965; Norekian and Moroz, 2019). We are only just starting to discover what the sensory cells and neural networks look like in other ctenophores (Norekian and Moroz, 2019, 2020). It is likely that a closer look at sponge tissues will also reveal a diverse set of sensory cells and also common mechanisms of signalling. Behaviors seen in deep sea glass sponges (Kahn et al., 2020), as well as those in shallow water demosponges (Nickel, 2004; Leys et al., 2019; Goldstein et al., 2020), are diverse and very likely are coordinated by sensory-receptor and "integrator" cell types that we are still unfamiliar with. Given the diversity and long evolutionary history of sponges, it can be expected that many different cell types carry out sense and information transmission functions. The richness of such a diverse set of "sensory/neural-type" cells might explain why many poriferans have almost twice as many genes as most animals with neurons (Fernandez-Valverde et al., 2015; Kenny et al., 2020).

6.8 CONCLUSIONS AND FUTURE DIRECTIONS

Much of the above discussion has focused on looking for common neurotransmitter types, transcriptional signatures, or synaptic structures and has come up with little to indicate that there is a shared sensory-neural cell type. There are many different ways in which behavior can be coordinated in metazoans, and this probably reflects independent origins from multiple different cell types with distinct transcriptomic signatures. For example, although it is suggested that paracrine communication between gland cells is involved in the concerted ciliary behavior that occurs during feeding in *Trichoplax* (Mayorova et al., 2019), mathematical models have also illustrated that the movement of *Trichoplax* towards food could result from the collective actions of ciliated cells acting independently (Smith et al., 2019). The sponge larval photoreceptor cells are also thought to respond individually to light with the collective responses of all cells individually causing the larva to steer towards or away from the light (Leys and Degnan, 2001; Collin et al., 2010). Even larval bilaterians have unusual non-neuronal methods of coordination. The gastrulating larva of the brachiopod *Terebratalia transversa* expresses opsin genes and can respond to light even before neurons are differentiated (Passamaneck et al., 2011). Glass sponge syncytia are electrically excitable epithelia that conduct calcium potentials all the way through to collar flagella units triggering flagella arrest (Mackie et al., 1983; Leys et al., 2007). The fact that syncytia can propagate electrical signals resulting in the coordinated arrest of the feeding current in a large animal (glass sponges can easily be a meter tall) illustrates that evolution has produced different cell types that are able to function in different ways but in a manner analogous to neurons.

There is some evidence that transcription factors involved in the development of sensory-neural tissues in bilaterians are expressed in sensory cells in non-bilaterians (Arendt, 2003; Richards et al., 2008; Arendt et al., 2016). But there is no common neuronal molecular signature that characterizes cells of a "pre-neural" fate in animals without neurons (Sebé-Pedrós et al., 2018a). Since this is the case, one new direction we think would be useful to illuminate origins of the neuronal cell type is to "trace forward" the biology of neurons of non-bilaterians, that is, to find the core set of genes shared by these non-bilaterian neurons and explore how those genetic modules have been propagated through the tree as evolutionary time has progressed.

As Liebeskind et al. (2016) point out, "phenotypes and their molecular constituents have become unlinked over deep evolutionary time," and so there is little convergence between the morphology and molecular signature of the neuronal phenotype. The sensory-neural cell type is wonderfully diverse and has repeatedly evolved similar characteristics in many different lineages of metazoans, and only within lineages has it become canalized both in form and transcriptomic signature (Fernández and Gabaldón, 2020). Another way forward is therefore to explore in depth the neuronal phenotype in each lineage of non-bilaterian, following the lead of new work on placozoans, which has expanded knowledge of cell morphology (Smith et al., 2014), physiology of ion channels (Senatore et al., 2017), behavior and communication (Varoqueaux et al., 2018), and expression profiles of individual cells (Sebé-Pedrós et al., 2018a). Integrating that data with single cell ATAC-seq and protein data (e.g., Stuart et al., 2019) could provide a more holistic view of the sensory or neuronal or signalling cell types.

ACKNOWLEDGEMENTS

SPL acknowledges funding from the National Science and Engineering Research Council (Canada) through a Discovery Grant and from the Yale Systems Biology Institute (via G. Wagner) to attend the Division of Evolutionary and Developmental Biology Workshop on Animal Cell Types: Their Origin and Evolution at the Society for Integrative and Comparative Biology, San Francisco, 2018 where some of this work was presented.

REFERENCES

Achim K, Arendt D (2014) Structural evolution of cell types by step-wise assembly of cellular modules. *Curr Opin Genet Dev.* 27:102–108.
Achim K, Pettit J-B, Saraiva LR, Gavriouchkina D, Larsson T, Arendt D, Marioni JC (2015) High-throughput spatial mapping of single-cell RNA-seq data to tissue of origin. *Nat Biotechnol.* 33:503–509.
Anctil M (2015) *The Dawn of the Neuron*. McGill-Queens University Press, Montreal.
Anderson P, Grünert U (1988) Three-dimensional structure of bidirectional, excitatory chemical synapses in the jellyfish *Cyanea capillata*. *Synapse.* 2:603–613.
Anderson P, Schwab WE (1981) The organization and structure of nerve and muscle in the jellyfish *Cyanea capillata* (Coelenterata; Scyphozoa). *J Morphol.* 170:383–399.
Anderson PAV (2015) On the origins of that most transformative of biological systems – The nervous system. *J Exp Biol.* 218:504–505.
Arendt D (2003) Evolution of eyes and photoreceptor cells. *Int J Dev Biol.* 47:563–571.
Arendt D (2020) The evolutionary assembly of neuronal machinery. *Curr Biol.* 30:R603–R616.
Arendt D, Musser JM, Baker CVH, Bergman A, Cepko C, Erwin DH, Pavlicev M, Schlosser G, Widder S, Laubichler MD, Wagner GP (2016) The origin and evolution of cell types. *Nat Rev Genet.* 17:744–757.
Assmann M, Kuhn A, Dürrnagel S, Holstein TW, Gründer S (2014) The comprehensive analysis of DEG/ENaC subunits in Hydra reveals a large variety of peptide-gated channels, potentially involved in neuromuscular transmission. *BMC Biol.* 12:84–84.
Baran Y, Bercovich A, Sebe-Pedros A, Lubling Y, Giladi A, Chomsky E, Meir Z, Hoichman M, Lifshitz A, Tanay A (2019) MetaCell: analysis of single-cell RNA-seq data using K-nn graph partitions. *Genome Biol.* 20:206.
Bidder GP (1937) The perfection of sponges. *Proc Linn Soc Lond.* 149:119–146.
Boeckers TM (2006) The postsynaptic density. *Cell Tiss Res.* 326:409–422.
Bone Q (2005) Gelatinous animals and physiology. *JMBA.* 85:641–653.
Boury-Esnault N, Rützler K (1997) Thesaurus of sponge morphology. *Smithson Contrib Zool.* 596:1–55.
Bucher D, Anderson PAV (2015) Evolution of the first nervous systems – What can we surmise? *J Exp Biol.* 218:501.
Burkhardt P (2015) The origin and evolution of synaptic proteins – Choanoflagellates lead the way. *J Exp Biol.* 218:506–514.
Cao J, Packer JS, Ramani V, Cusanovich DA, Huynh C, Daza R, Qiu X, Lee C, Furlan SN, Steemers FJ, Adey A, Waterston RH, Trapnell C, Shendure J (2017) Comprehensive single-cell transcriptional profiling of a multicellular organism. *Science.* 357:661–667.
Chun C (1880) Die ctenophoren des golfes von neapel. Fauna flora neapel. *MonogDB* 1:xviii + 313 p. 318 Taf.
Claus C (1878) Studien über polypen und quallen der adria. *Denkschr Akad Wiss Wien* Bd. 38:1–64, Taf. 61–11.
Collin R, Mobley AS, Busutil Lopez L, Leys SP, Cristina Diaz M, Thacker RW (2010) Phototactic responses of larvae from the marine sponges *Neopetrosia proxima* and *Xestospongia bocatorensis* (Haplosclerida: Petrosiidae). *Invertebr Biol.* 129:121–128.
Croset V, Treiber CD, Waddell S (2018) Cellular diversity in the *Drosophila* midbrain revealed by single-cell transcriptomics. *eLife* 7:e34550.

DuBuc TQ, Ryan JF, Martindale MQ (2019) "Dorsal–ventral" genes are part of an ancient axial patterning system: Evidence from *Trichoplax adhaerens* (Placozoa). *Mol Biol Evol*. 36:966–973.

Duménieu M, Oulé M, Kreutz MR, Lopez-Rojas J (2017) The segregated expression of voltage-gated potassium and sodium channels in neuronal membranes: Functional implications and regulatory mechanisms. *Front Cell Neurosci*. 11:115.

Dunn CW, Hejnol A, Matus DQ, Pang K, Browne WE, Smith SA, Seaver E, Rouse GW, Obst M, Edgecombe GD, Sorensen MV, Haddock SHD, Schmidt-Rhaesa A, Okusu A, Mobjerg Kristensen R, Wheeler WC, Martindale MQ, Giribet G (2008) Broad phylogenomic sampling improves resolution of the animal tree of life. *Nature*. 452:745–749.

Dunn CW, Leys SP, Haddock SHD (2015) The hidden biology of sponges and ctenophores. *Trends Ecol Evol* 30:282–291.

Elliott GRD (2009) *The Cytology and Physiology of Coordinated Behaviour in a Freshwater Sponge, Ephydatia muelleri*. PhD Thesis, University of Alberta.

Faltine-Gonzalez DZ, Layden MJ (2019) Characterization of nAChRs in *Nematostella vectensis* supports neuronal and non-neuronal roles in the cnidarian-bilaterian common ancestor. *Evodevo*. 10:27–27.

Farquhar MG, Palade GE (1963) Junctional complexes in various epithelia. *J Cell Biol*. 17:375–412.

Fernandez-Valverde SL, Calcino AD, Degnan BM (2015) Deep developmental transcriptome sequencing uncovers numerous new genes and enhances gene annotation in the sponge *Amphimedon queenslandica*. *BMC Genom*. 16:387–387.

Fernández R, Gabaldón T (2020) Gene gain and loss across the metazoan tree of life. *Nat Ecol Evol*. 4:524–533.

Feuda R, Dohrmann M, Pett W, Philippe H, Rota-Stabelli O, Lartillot N, Wörheide G, Pisani D (2017) Improved modeling of compositional heterogeneity supports sponges as sister to all other animals. *Curr Biol*. 27:3864–3870.e3864.

Fortunato S, Adamski M, Bergum B, Guder C, Jordal S, Leininger S, Zwafink C, Rapp HT, Adamska M (2012) Genome-wide analysis of the sox family in the calcareous sponge *Sycon ciliatum*: Multiple genes with unique expression patterns. *EvoDevo*. 3:14.

Fortunato SAV, Leininger S, Adamska M (2014) Evolution of the pax-six-eya-dach network: the calcisponge case study. *EvoDevo*. 5:23.

Fortunato SAV, Vervoort M, Adamski M, Adamska M (2016) Conservation and divergence of bHLH genes in the calcisponge *Sycon ciliatum*. *EvoDevo*. 7:23.

Galliot B, Quiquand M, Ghila L, de Rosa R, Miljkovic-Licina M, Chera S (2009) Origins of neurogenesis, a cnidarian view. *Dev Biol*. 332:2–24.

Garm A, Andersson F, Nilsson D-E (2008) Unique structure and optics of the lesser eyes of the box jellyfish *Tripedalia cystophora*. *Vision Res*. 48:1061–1073.

Germain G, Anctil M (1996) Evidence for intercellular coupling and connexin-like protein in the luminescent endoderm of *Renilla koellikeri* (Cnidaria, Anthozoa). *Biol Bull*. 191:353–366.

Goldstein J, Bisbo N, Funch P, Riisgård HU (2020) Contraction-expansion and the effects on the aquiferous system in the demosponge *Halichondria panicea*. *Front Mar Sci*. 7:113.

Golz R, Thurm U (1994) The ciliated sensory cell of *Stauridiosarsia producta* (Cnidaria, Hydrozoa)—A nematocyst-free nematocyte? *Zoomorphology* 114:185–194.

Grimmelikhuijzen C, Westfall JA (1995) The nervous systems of Cnidarians. In: Breidbach O, Kutsch W (eds) *The Nervous Systems of Invertebrates: An Evolutionary and Comparative Approach*. Birkhauser Verlag, Basel, Switzerland, pp. 7–24.

Gründer S, Assmann M (2015) Peptide-gated ion channels and the simple nervous system of *Hydra*. *J Exp Biol*. 218:551.

Hall C, Rodriguez M, Garcia J, Posfai D, DuMez R, Wictor E, Quintero OA, Hill MS, Rivera AS, Hill AL (2019) Secreted frizzled related protein is a target of PaxB and plays a role in aquiferous system development in the freshwater sponge, *Ephydatia muelleri*. *PLoS One*. 14:e0212005.

Hartenstein V (2016) Platyhelminthes (excluding neodermata). In: Schmidt-Rhaesa A, Harzsch S, Purschke G (eds) *Structure and Evolution of Invertebrate Nervous Systems*. Oxford University Press, Oxford, pp. 74–92.

Hartenstein V, Stollewerk A (2015) The evolution of early neurogenesis. *Dev Cell*. 32:390–407.

Hartwell LH, Hopfield JJ, Leibler S, Murray AW (1999) From molecular to modular cell biology. *Nature* 402 Supplement:C47–C52.

Hejnol A (2016) Acoelomorpha. In: Schmidt-Rhaesa A, Harzsch S, Purschke G (eds) *Structure and Evolution of Invertebrate Nervous Systems*. Oxford University Press, Oxford, pp. 56–61.

Hernandez-Nicaise M-L (1973) The nervous system of ctenophores III. Ultrastructure of synapses. *J Neurocytol*. 2:249–263.

Hernandez-Nicaise M.-L. (1974) Ultrastructural evidence for a sensory-motor neuron in Ctenophora. *Tiss Cell*. 6:43–47.

Hernandez-Nicaise M-L, Mackie GO, Meech RW (1980) Giant smooth muscle cells of *Beroë*: Ultrastructure, innervation, and electrical properties. *J Gen Physiol*. 75:79–105.

Hertwig O (1880) Uber den bau den ctenophoren. *Jenaisiche Z Naturwiss*. 14:393–457.

Hertwig O, Hertwig R (1878) *Das Nervensystem und die Sinnesorgane der Medusen (The Nervous System and the Sensory Organs of the Medusa)*. Vogel, Leipzig.

Hill A, Boll W, Ries C, Warner L, Osswalt M, Hill M, Noll M (2010) Origin of pax and six gene families in sponges: Single PaxB and Six1/2 orthologs in *Chalinula loosanoffi*. *Dev Biol*. 343:106–123.

Horridge GA (1965) Non-motile sensory cilia and neuromuscular junctions in a ctenophore independent effector organ. *Proc R Soc Lond B*. 162:333–350.

Hoshiyama D, Iwabe N, Miyata T (2007) Evolution of the gene families forming the Pax/Six regulatory network: Isolation of genes from primitive animals and molecular phylogenetic analyses. *FEBS Lett*. 581:1639–1643.

Hoshiyama D, Suga H, Iwabe N, Koyanagi M, Nikoh N, Kuma K, Matsuda F, Honjo T, Miyata T (1998) Sponge *Pax* cDNA related to *Pax* 2-5-8 and ancient gene duplications in the *Pax* family. *J Mol Evol*. 47:640–648.

Huang ZJ, Paul A (2019) The diversity of GABAergic neurons and neural communication elements. *Nat Rev Neurosci*. 20:563–572.

Jacobs D, Nakanishi N, Yuan D, Camara A, Nichols S, Hartenstein V (2007) Evolution of sensory structures in basal metazoa. *Integr Comp Biol*. 47:712–723.

Jager M, Quéinnec E, Le Guyader H, Manuel M (2011) Multiple Sox genes are expressed in stem cells or in differentiating neuro-sensory cells in the hydrozoan *Clytia hemisphaerica*. *EvoDevo*. 2:12.

Jones WC (1962) Is there a nervous system in sponges? *Biol Rev Camb Philos Soc*. 37:1–50.

Kahn AS, Pennelly CW, McGill PR, Leys SP (2020) Behaviors of sessile benthic animals in the abyssal northeast Pacific Ocean. *Deep Sea Res II*. 173:104729

Kenny NJ, Francis WR, Rivera-Vicéns RE, Juravel K, de Mendoza A, Díez-Vives C, Lister R, Bezares-Calderon L, Grombacher L, Roller M, Barlow LD, Camilli S, Ryan JF, Wörheide G, Hill AL, Riesgo A, Leys SP (2020) Tracing animal genomic evolution with the chromosomal-level assembly of the freshwater sponge *Ephydatia muelleri*. *Nat Commun*. 11. doi:10.1038/s41467-41020-17397-w.

Kleinenberg N (1872) *Hydra – Eine anatomischentwicklungsgeschichtliche untersuchung (An anatomical-evolutionary investigation of Hydra)*. Wilhelm Engelmann, Leipzig.

Lanna E, Klautau M (2012) Embryogenesis and larval ultrastructure in *Paraleucilla magna* (Calcarea, Calcaronea), with remarks on the epilarval trophocyte epithelium ("placental membrane"). *Zoomorphogy* 131:277–292.

Larroux C, Luke GN, Koopman P, Rokhsar DS, Shimeld SM, Degnan BM (2008) Genesis and expansion of metazoan transcription factor gene classes. *Mol Biol Evol*. 25:980–996.

Layden M, Boekhout M, Martindale M (2012) *Nematostella vectensis* achaete-scute homolog NvashA regulates embryonic ectodermal neurogenesis and represents an ancient component of the metazoan neural specification pathway. *Development*. 139:1013–1022.

Leterrier C (2018) The axon initial segment: An updated viewpoint. *J Neurosci*. 38:2135–2145.

Leys SP (2007) Sponge coordination, tissues, and the evolution of gastrulation. In: Custodio M, Lobo-Hadju G, Lobo-Hadju E, Muricy G (eds) *Porifera Research: Biodiversity, Innovation and Sustainabiity*. 7th International Porifera Congress, Vol. 7. Museu Nacional, Rio de Janeiro, pp. 53–59.

Leys SP, Cronin T, Degnan BM, Marshall J (2002) Spectral sensitivity in a sponge larva. *J Comp Physiol A*. 188:199–202.

Leys SP, Degnan BM (2001) Cytological basis of photoresponsive behaviour in a sponge larva. *Biol Bull*. 201:323–338.

Leys SP, Mackie GO, Meech R (1999) Impulse conduction in a sponge. *J Exp Biol*. 202:1139–1150.

Leys SP, Mackie GO, Reiswig H (2007) The biology of glass sponges. *Adv Mar Biol*. 52:1–145.

Leys SP (1995) Cytoskeletal architecture and organelle transport in giant syncytia formed by fusion of hexactinellid sponge tissues. *Biol Bull*. 188:241–254.

Leys SP, Mah JL, McGill PR, Hamonic L, De Leo FC, Kahn AS (2019) Sponge behavior and the chemical basis of responses: A post-genomic view. *Integr Comp Biol*. 59:751–764.

Liebeskind BJ, Hillis DM, Zakon HH (2015) Convergence of ion channel genome content in early animal evolution. *PNAS*. 112:E846.

Liebeskind BJ, Hillis DM, Zakon HH, Hofmann HA (2016) Complex homology and the evolution of nervous systems. *TREE*. 31:127–135.

Liebeskind BJ, Hofmann HA, Hillis DM, Zakon HH (2017) Evolution of animal neural systems. *Annu Rev Ecol Evol Syst*.48:377–398.

Lorenzo R, Onizuka M, Defrance M, Laurent P (2019) Combining single-cell RNA-sequencing with a molecular atlas unveils new markers for *C. elegans* neuron classes. *BioRxiv*. 826560.

Ludeman DA, Farrar N, Riesgo A, Paps J, Leys SP (2014) Evolutionary origins of sensation in metazoans: Functional evidence for a new sensory organ in sponges. *BMC Evol Biol*. 14, 3.

Mackie GO (1970) Neuroid conduction and the evolution of conducting tissues. *Quart Rev Biol*. 45:319–332.

Mackie GO (2004) Epithelial conduction: Recent findings, old questions, and where do we go from here? *Hydrobiologia*. 530–531:73–80.

Mackie GO, Lawn ID, Pavans de Ceccatty M (1983) Studies on hexactinellid sponges. II. Excitability, conduction and coordination of responses in *Rhabdocalyptus dawsoni* (Lambe 1873). *Philos Trans R Soc Lond Biol Sci* 301:401–418.

Mackie GO, Marx RM, Meech RW (2003) Central circuitry in the jellyfish *Aglantha digitale* IV. Pathways coordinating feeding behaviour. *J Exp Biol*. 206:2487.

Mackie GO, Singla CL (1983) Studies on hexactinellid sponges. I Histology of *Rhabdocalyptus dawsoni* (Lambe, 1873). *Philos Trans Roy Soc Lond B*. 301:365–400.

Maldonado M, Durfort M, McCarthy DA, Young CM (2003) The cellular basis of photobehavior in the tufted parenchymella larva of demosponges. *Mar Biol (Berl)*. 143:427–441.

Marlow H, Srivastava M, Matus D, Rokhsar D, Martindale M (2009) Anatomy and development of the nervous system of *Nematostella vectensis*, an anthozoan cnidarian. *Dev Neurobiol*. 69:235–254.

Mayorova TD, Hammar K, Winters CA, Reese TS, Smith CL (2019) The ventral epithelium of *Trichoplax adhaerens* deploys in distinct patterns cells that secrete digestive enzymes, mucus or diverse neuropeptides. *Biol Open.* 8:bio045674.

McIntire SL, Reimer RJ, Schuske K, Edwards RH, Jorgensen EM (1997) Identification and characterization of the vesicular GABA transporter. *Nature.* 389:870–876.

Meech RW (2017) The evolution of neurons. In: Shepherd S (ed) *The Wiley Handbook of Evolutionary Neuroscience.* John Wiley & Sons, pp. 88–124.

Moroz LL (2009) On the independent origins of complex brains and neurons. *Brain Behav Evol.* 74:177–190.

Moroz LL (2015) Convergent evolution of neural systems in ctenophores. *J Exp Biol.* 218:598–611.

Moroz LL, Kocot KM, Citarella MR, Dosung S, Norekian TP, Povolotskaya IS, Grigorenko AP, Dailey C, Berezikov E, Buckley KM, Ptitsyn A, Reshetov D, Mukherjee K, Moroz TP, Bobkova Y, Yu F, Kapitonov VV, Jurka J, Bobkov YV, Swore JJ, Girardo DO, Fodor A, Gusev F, Sanford R, Bruders R, Kittler E, Mills CE, Rast JP, Derelle R, Solovyev VV, Kondrashov FA, Swalla BJ, Sweedler JV, Rogaev EI, Halanych KM, Kohn AB (2014) The ctenophore genome and the evolutionary origins of neural systems. *Nature.* 510:109–114.

Moroz LL, Kohn AB (2015) Unbiased view of synaptic and neuronal gene complement in ctenophores: Are there pan-neuronal and pan-synaptic genes across metazoa? *Integr Comp Biol.* 55:1028–1049.

Moroz LL, Kohn AB (2016) Independent origins of neurons and synapses: Insights from ctenophores. *Philos Trans Roy Soc B Biol Sci.* 371:20150041.

Nakanishi N, Renfer E, Technau U, Rentzsch F (2012) Nervous systems of the sea anemone *Nematostella vectensis* are generated by ectoderm and endoderm and shaped by distinct mechanisms. *Development.* 139:347–357.

Nakanishi N, Stoupin D, Degnan SM, Degnan BM (2015) Sensory flask cells in sponge larvae regulate metamorphosis via calcium signaling. *Integr Comp Biol.* 55:1018–1027.

Niciu MJ, Kelmendi B, Sanacora G (2012) Overview of glutamatergic neurotransmission in the nervous system. *Pharmacol Biochem Behav.* 100:656–664.

Nickel M (2004) Kinetics and rhythm of body contractions in the sponge *Tethya wilhelma* (Porifera: Demospongiae). *J Exp Biol.* 207:4515–4524.

Nickel M (2010) Evolutionary emergence of synaptic nervous systems: what can we learn from the non-synaptic, nerveless Porifera? *Invertebr Biol.* 129:1–16.

Nilsson D-E, Gislén L, Coates MM, Skogh C, Garm A (2005) Advanced optics in a jellyfish eye. *Nature.* 435:201.

Niswender CM, Conn PJ (2010) Metabotropic glutamate receptors: physiology, pharmacology, and disease. *Annu Rev Pharmacol Toxicol.* 50:295–322.

Norekian TP, Moroz LL (2019) Neural system and receptor diversity in the ctenophore *Beroe abyssicola*. *J Comp Neurol.* 527:1986–2008.

Norekian TP, Moroz LL (2020) Comparative neuroanatomy of ctenophores: Neural and muscular systems in *Euplokamis dunlapae* and related species. *J Comp Neurol.* 528:481–501.

Palade GE, Palay S (1954) Electron microscope observations of interneural and neuromuscular synapses. *Anatom Record.* 118:335.

Parker GH (1911) The origin and significance of the primitive nervous system. *Proc Amer Philos Soc.* 50:217–225.

Parker GH (1919) The elementary nervous system. *Monogr. Exp Biol.* 30:25–49.

Passamaneck YJ, Furchheim N, Hejnol A, Martindale MQ, Lüter C (2011) Ciliary photoreceptors in the cerebral eyes of a protostome larva. *EvoDevo.* 2:6.

Passano LM (1982) Scyphozoa and cubozoa. In: Shelton GAB (ed) *Electrical Conduction and Behaviour in 'Simple' Invertebrates*. Clarenden Press, Oxford, pp 149–242.

Paul A, Crow M, Raudales R, He M, Gillis J, Huang ZJ (2017) Transcriptional architecture of synaptic communication delineates GABAergic neuron identity. *Cell*. 171:522–539.e520.

Pavans de Ceccatty M (1955) Le systeme nerveux des éponges calcaires et siliceuses. *Ann Sci Nat Zool*. 17:203–288.

Pavans de Ceccatty M (1974) Coordination in sponges. The foundations of integration. *Am Zool*. 14:895–903.

Pavans de Ceccatty M (1979) Cell correlations and integration in Sponges. *Colloq Int CNRS biol Spongiaires*. 291:123–135.

Pavans de Ceccatty M, Thiney Y, Garrone R (1970) Les bases ultrastructurales des communications intercellulaires dans les oscules de quelques éponges. In: Fry W (ed) *The Biology of the Porifera (Zoological Society of London)*. Academic Press, London, pp. 449–466.

Pett W, Adamski M, Adamska M, Francis WR, Eitel M, Pisani D, Wörheide G (2019) The role of homology and orthology in the phylogenomic analysis of metazoan gene content. *Mol Biol Evol*. 36:643–649.

Ramon y Cajal S (1899) *Histology of the Nervous System*. Oxford University Press, New York, NY.

Richards GS, Degnan BM (2012) The expression of delta ligands in the sponge *Amphimedon queenslandica* suggests an ancient role for notch signaling in metazoan development. *EvoDevo*. 3:15.

Richards GS, Simionata E, Perron M, Adamska M, Vervoort M, Degnan BM (2008) Sponge genes provide new insight into the evolutionary origin of the neurogenic circuit. *Curr Biol*. 18:1156–1161.

Rivera A, Winters I, Rued A, Ding S, Posfai D, Cieniewicz B, Cameron K, Gentile L, Hill A (2013) The evolution and function of the Pax/Six regulatory network in sponges. *Evol Dev*. 15:186–196.

Rolls MM, Jegla TJ (2015) Neuronal polarity: An evolutionary perspective. *J Exp Biol*. 218:572.

Ryan JF, Chiodin M (2015) Where is my mind? How sponges and placozoans may have lost neural cell types. *Philos Trans R Soc Lond B*. 370:20150059.

Ryan JF, Pang K, Mullikin JC, Martindale MQ, Baxevanis AD, Program NCS (2010) The homeodomain complement of the ctenophore *Mnemiopsis leidyi* suggests that Ctenophora and Porifera diverged prior to the ParaHoxozoa. *EvoDevo*. 1:9.

Ryan JF, Pang K, Schnitzler CE, Nguyen A-D, Moreland RT, Simmons DK, Koch BJ, Francis WR, Havlak P, Smith SA, Putnam NH, Haddock SHD, Dunn CW, Wolfsberg TG, Mullikin JC, Martindale MQ, Baxevanis AD (2013) The genome of the ctenophore *Mnemiopsis leidyi* and its implications for cell type evolution. *Science*. 342:1242592.

Satterlie R (2019) *Cnidarian Neurobiology The Oxford Handbook of Invertebrate Neurobiology*. Oxford University Press, Oxford.

Satterlie R, Spencer A (1987) Organization of conducting systems in "simple" invertebrates: Porifera, Cnidaria and Ctenophora. In: Ali M (ed) *Nervous Systems in Invertebrates*. Springer, Boston, MA, pp. 213–264.

Schaefer E (1900) *Text Book of Physiology*. Young J. Pentland, Edinburgh and London.

Schnitzler CE, Simmons DK, Pang K, Martindale MQ, Baxevanis AD (2014) Expression of multiple Sox genes through embryonic development in the ctenophore *Mnemiopsis leidyi* is spatially restricted to zones of cell proliferation. *EvoDevo*. 5:15.

Sebé-Pedrós A, Chomsky E, Pang K, Lara-Astiaso D, Gaiti F, Mukamel Z, Amit I, Hejnol A, Degnan BM, Tanay A (2018a) Early metazoan cell type diversity and the evolution of multicellular gene regulation. *Nat Ecol Evol*. 2:1176–1188.

Sebé-Pedrós A, Saudemont B, Chomsky E, Plessier F, Mailhé M-P, Renno J, Loe-Mie Y, Lifshitz A, Mukamel Z, Schmutz S, Novault S, Steinmetz PRH, Spitz F, Tanay A, Marlow H (2018b) Cnidarian cell type diversity and regulation revealed by whole-organism single-cell RNA-Seq. *Cell.* 173:1520–1534.e1520.

Senatore A, Reese TS, Smith CL (2017) Neuropeptidergic integration of behavior in *Trichoplax adhaerens*, an animal without synapses. *J Exp Biol.* 220:3381–3390.

Shelton GAB (1982) Anthozoa. In: Shelton GAB (ed) *Electrical Conduction and Behaviour in 'Simple' Invertebrates.* Clarenden Press, Oxford, pp. 203–242.

Shubin N, Tabin C, Carroll S (2009) Deep homology and the origins of evolutionary novelty. *Nature* 457:818–823.

Simmons D, Pang K, Martindale MQ (2012) Lim homeobox genes in the ctenophore mnemiopsis leidyi: the evolution of neural cell type specification. *EvoDevo.* 3:2.

Simpson TL (1984) *The Cell Biology of Sponges.* Springer Verlag, New York, NY.

Singla CL (1974) Ocelli of hydromedusae. *Cell Tissue Res.* 149:413–429.

Smith CL, Reese T, Govezensky T, Barrio R (2019) Coherent directed movement toward food modeled in *Trichoplax*, a ciliated animal lacking a nervous system. *PNAS.* 116:8901–8908.

Smith CL, Varoqueaux F, Kittelmann M, Azzam R, Cooper B, Winters C, Eitel M, Fasshauer D, Reese T (2014) Novel cell types, neurosecretory cells, and body plan of the early-diverging metazoan *Trichoplax adhaerens*. *Curr Biol.* 24:1565–1572.

Smith CL, Mayorova TD (2019) Insights into the evolution of digestive systems from studies of *Trichoplax adhaerens*. *Cell Tissue Res.* 377:353–367.

Smith CL, Pivovarova N, Reese TS (2015) Coordinated feeding behavior in *Trichoplax*, an animal without synapses. *PLoS One.* 10:e0136098.

Spencer AN, Schwab WE (1982) Hydrozoa. In: Shelton GAB (ed) *Electrical Conduction and Behaviour in 'Simple' Invertebrates.* Clarenden Press, Oxford.

Stuart T, Butler A, Hoffman P, Hafemeister C, Papalexi E, Mauck WM, Hao Y, Stoeckius M, Smibert P, Satija R (2019) Comprehensive integration of single-cell data. *Cell* 177:1888–1902.e1821.

Tamm S (1982) Ctenophora. In: Shelton GAB (ed) *Electrical Conduction and Behaviour in 'Simple' Invertebrates.* Clarenden Press, Oxford, pp. 266–358.

Telford MJ, Moroz LL, Halanych KM (2016) A sisterly dispute. *Nature* 529:286–287.

Torrealba F, Carrasco MA (2004) A review on electron microscopy and neurotransmitter systems. *Brain Res Rev.* 47:5–17.

Tuzet O (1970) La polarité de l'oeuf et la symétrie de la larve des éponges calcaires. *Zool Soc Lond.* 25:437–448.

Ueda N, Richards G, Degnan BM, Kranz A, Adamska M, Croll R, Degnan SM (2016) An ancient role for nitric oxide in regulating the animal pelagobenthic life cycle: Evidence from a marine sponge. *Sci Rep.* 6:37546.

Varoqueaux F, Williams EA, Grandemange S, Truscello L, Kamm K, Schierwater B, Jékely G, Fasshauer D (2018) High cell diversity and complex peptidergic signaling underlie placozoan behavior. *Curr Biol.* 28:3495–3501.e3492.

Voronezhskaya E, Croll R (2016) Mollusca: Gastropoda. In: Schmidt-Rhaesa A, Harzsch S, Purschke G (eds) *Structure and Evolution of Invertebrate Nervous Systems.* Oxford University Press, Oxford, pp. 196–221.

Waldeyer W (1891) *Uber einige neurer Forschungen im Gebiete der Anatomie des Centralnervensystems*, Vol. 17. Hansebooks, Norderstedt, 1213–1218; 1244–1216; 1267–1219; 1331–1212; 1352–1216.

Westfall J (1987) Ultrastructure of Invertebrate Synapses. In: Ali M (ed) *Nervous Systems in Invertebrates.* Springer, Boston, MA, pp. 3–28.

Whelan NV, Kocot KM, Moroz LL, Halanych KM (2015) Error, signal, and the placement of Ctenophora sister to all other animals. *PNAS*. 112:5773–5778.

Wojcik SM, Rhee JS, Herzog E, Sigler A, Jahn R, Takamori S, Brose N, Rosenmund C (2004) An essential role for vesicular glutamate transporter 1 (VGLUT1) in postnatal development and control of quantal size. *PNAS*. 101:7158–7163.

Zimmermann B, Robert NSM, Technau U, Simakov O (2019) Ancient animal genome architecture reflects cell type identities. *Nat Ecol Evol*. 3:1289–1293.

7 Cell Types, Morphology, and Evolution of Animal Excretory Organs

Carmen Andrikou, Ludwik Gąsiorowski, and Andreas Hejnol

CONTENTS

7.1 Introduction 129
7.2 Diversity of Cell Types in Excretory Organs: A Morphological
Perspective 131
    7.2.1 Secretory Excretory Organs (Malpighian Tubules and Others) 132
    7.2.2 Protonephridia 133
    7.2.3 Metanephridial Systems 136
    7.2.4 Excretory Cells 139
    7.2.5 Excretory Organs and Adaptations to New Environments................ 140
7.3 Molecular Identity of Excretory Organs: From Development to Function 140
    7.3.1 Nephron Development 141
    7.3.2 Protonephridial Development 143
    7.3.3 Development of Insect Malpighian Tubules 144
    7.3.4 Molecular Basis of Functional Compartmentalization..................... 144
    7.3.5 Nitrogenous Waste Excretion 147
7.4 The Problem of Homology 148
References.. 150

7.1 INTRODUCTION

All animals (Metazoa) need to excrete metabolic waste products from their bodies (Larsen et al., 2014; Schmidt-Nielsen, 1997; Schmidt-Rhaesa, 2007). On a cellular level, this process takes place via transmembrane proteins that are specialized for transporting these products in the context of ion gradients (Ichimura and Sakai, 2017; Larsen et al., 2014; O'Donnell, 2010; Schmidt-Nielsen, 1997; Weihrauch and O'Donnell, 2017). It is widely believed that non-bilaterian animals excrete via passive diffusion through their integument, although this hypothesis has been challenged by a recent work on cnidarian and xenacoelomorph species (Andrikou et al., 2019). However, most bilaterians possess specialized excretory organs that remove metabolites more efficiently (Figure 7.1). These organs are diverse, and their evolutionary

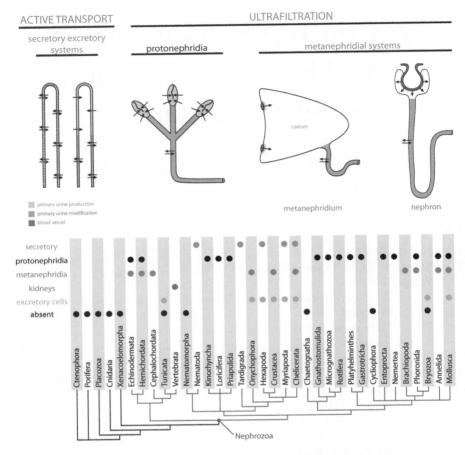

FIGURE 7.1 Excretory systems distribution across the animal tree. Schematic drawings and typology of basic types of excretory organs with color coded sites of primary urine production and modification (see text for details). Distribution of different types of excretory organs in nephrozoan clades.

relationship has puzzled many zoologists since their discovery (Bartolomaeus and Ax, 1992; Bartolomaeus and Quast, 2005; Goodrich, 1945; Ichimura and Sakai, 2017; Koch et al., 2014; Ruppert, 1994; Ruppert and Smith, 1988; Schmidt-Rhaesa, 2007). Nephridia, kidneys, Malpighian tubules etc. are composed out of different cell types that can be discriminated by their morphology and function (e.g. Goodrich, 1945; Ruppert and Smith, 1988; Schmidt-Rhaesa, 2007). The embryology of these organs is also intriguing, because it varies between species and also involves, in some cases (e.g. metanephridia), an interaction between cells of different germ layers, such as the mesoderm and ectoderm (Bartolomaeus, 1989; Goodrich, 1945; Lüter, 1995; Ruppert, 1994; Schmidt-Rhaesa, 2007). When and how many times these specialized organs evolved remains unclear. In the context of the placement of the Xenacoelomorpha as sister group to all remaining Bilateria, a new taxon name for Protostomia + Deuterostomia has been introduced, namely Nephrozoa (Jondelius et al., 2002). This refers to the presence of excretory organs (i.e. nephridia) in the last common ancestor of this clade that would also be an apomorphy of Nephrozoa (Figure 7.1). Here we aim to describe the astonishing variability seen in excretory organs from a cell-type perspective, for which the diversity in morphology, development, and functional composition can be particularly challenging. We interpret this diversity from an evolutionary perspective and discuss problems in homologization on different levels.

7.2 DIVERSITY OF CELL TYPES IN EXCRETORY ORGANS: A MORPHOLOGICAL PERSPECTIVE

Excretion in most of the Nephrozoa is a two-stage process (Ichimura and Sakai, 2017; Ruppert, 1994; Ruppert and Smith, 1988; Schmidt-Nielsen, 1997; Schmidt-Rhaesa, 2007). Initially, the body fluid (e.g. blood, haemolymph, or interstitial fluid) is roughly filtered from large proteins and cells to produce the so-called primary urine (Ichimura and Sakai, 2017; Schmidt-Rhaesa, 2007). This initial product becomes later secondarily modified (e.g. the ion concentration or water volume can be specifically changed), resulting in the finite secondary urine, which is eventually expelled from the body (Schmidt-Nielsen, 1997; Schmidt-Rhaesa, 2007). As there are many ways in which those two processes can be performed, the excretory organs vary greatly regarding their morphology and physiology and, as a consequence, in the diversity and spatial arrangement of the particular cell types, which built them.

Traditionally, zoologists group excretory organs into categories based on their physiological and structural properties, which reflect functional rather than evolutionary similarities (Figure 7.1; Bartolomaeus and Ax, 1992; Ichimura and Sakai, 2017; Ruppert and Smith, 1988; Schmidt-Rhaesa, 2007). The most basic division relates to the mechanism of primary urine production—in secretory excretory organs, active intracellular transport is used for that purpose, whereas in ultrafiltration-based systems, the primary urine is filtered through a semipermeable extracellular membrane—a filter composed of either extracellular matrix (ECM) or slit diaphragm or both (Figure 7.1; Ichimura and Sakai, 2017; Schmidt-Nielsen, 1997; Schmidt-Rhaesa, 2007). The latter category includes protonephridia, where ultrafiltration is driven by the ciliary action, and the metanephridial system, in which blood pressure is used to produce primary urine (Bartolomaeus and Ax, 1992; Ichimura and Sakai, 2017; Ruppert, 1994; Ruppert and Smith, 1988; Schmidt-Rhaesa, 2007). Among metanephridial systems, it is possible to distinguish between vertebrate kidneys, in which ultrafiltration and secondary urine modification occurs in the single structural unit (i.e. nephron with glomerulus) and invertebrate metanephridial systems, in which ultrafiltration and secondary urine modification are spatially separated, the former happening in the coelom lining and the latter in the metanephridium itself (Ichimura and Sakai, 2017; Schmidt-Nielsen, 1997; Schmidt-Rhaesa, 2007).

7.2.1 SECRETORY EXCRETORY ORGANS (MALPIGHIAN TUBULES AND OTHERS)

Among the best-known examples of the secretory excretory organs are the Malpighian tubules, the prevalent excretory organs in many Panarthropoda taxa. Although Malpighian tubules are found in representatives of Eutardigrada, Chelicerata, Myriapoda, and Hexapoda (Figure 7.1), it remains unknown if they are all homologous to each other or rather if they evolved independently in particular lineages (Bitsch and Bitsch, 2004; Greven, 1982; Paulus, 2000). Those organs are tubular invaginations of the gut (originating close to the midgut-hindgut transition), which penetrate the haemocoel; distally, they are blindly ended, and proximally, they open to the gut lumen (Berridge and Oschman, 1969; Eichelberg and Wessing, 1975; Nocelli et al., 2016; Schmidt-Nielsen, 1997; Schmidt-Rhaesa, 2007). Malpighian tubules are surrounded by a basal lamina, which serves as a barrier for cells and large proteins. Both their external and internal surface is greatly increased by basal infoldings and microvillar structures respectively, which enhance cellular uptake and secretion (Berridge and Oschman, 1969; Maddrell, 1980). In insects, the organs are additionally equipped with contractile muscle fibers and tracheae (e.g. Bradley, 1983; Garayoa et al., 1992; Li et al., 2015; Taylor, 1971b; Wall et al., 1975).

Often, there are two basic cell types in Malpighian tubules—e.g. many insects possess 1) primary cells (also known as type I cells), which produce primary urine and are characterized by numerous mitochondria-rich, external, basal infoldings and internal microvilli and 2) less abundant secondary or stellate cells (also known as type II cells) with less developed microvilli, which are likely involved in the secondary urine modification (e.g. Berridge and Oschman, 1969; Dow and Davies, 2001; Kapoor, 1994; Li et al., 2015; Nocelli et al., 2016; Pal and Kumar, 2012; Taylor, 1971a,b; Wall et al., 1975). In some animals, Malpighian tubules seem to include additional cell types—e.g. in tardigrades, "supportive" cells are present next to primary and secondary cells (Møbjerg and Dahl, 1996; Węglarska, 1980).

It is evident that not only cell type diversity, but also their spatial distribution along the organ is crucial for the evolution of Malpighian tubules. The two basic cell types—the primary urine producing ones and the secondarily modifying ones—can be a) uniformly distributed along the tubule, as is case in many insects (Berridge and Oschman, 1969; Dow and Davies, 2001; Kapoor, 1994; Nocelli et al., 2016; Pal and Kumar, 2012; Taylor, 1971a,b; Wall et al., 1975), or b) restricted to the respectively proximal and distal portion of the tubule, resulting in a clear division of the organ into distinct parts responsible for urine production and secondary modification, as in tardigrades (Møbjerg and Dahl, 1996; Schill, 2019; Węglarska, 1980), millipedes (Johnson and Riegel, 1977), or some insects (Arab and Caetano, 2002; Bradley, 1983; Green, 1979, 1980; Li et al., 2015; Nicholls, 1983; Nocelli et al., 2016). The subdivision of Malpighian tubules into distinct regions with different cell types has been also shown in spiders; however, the function of the cells in each section is not clear (Hazelton et al., 2002; Seitz, 1987).

Secretory excretory organs are also found among nematodes in which the two different basic types can be recognized (reviewed in Chitwood and Chitwood, 1950). Some of the free-living, mostly marine forms (once united into the group

Aphasmidia=Adenophorea) have a single large glandular cell used for excretion, the so-called ventral gland, which opens to the exterior by a single pore. Members of the clade Phasmidia (=Secernentea), which include most of the parasitic forms, as well as a model species *Caenorhabditis elegans*, possess a specialized excretory organ, which consists of several cells, each with a different ultrastructure and function (e.g. McLaren, 1974; Nelson et al., 1983; Waddell, 1968). The exact number and types of cells might vary, but generally the organ always represents some modification of the so-called H-system (Chitwood and Chitwood, 1950). There is an unpaired sinus cell which opens to the exterior by a ventromedian terminal duct lined with cuticle. From the sinus, two pairs of longitudinal, lateral canals extend in the lateral hypodermal cords to the anterior and posterior portion of the animal. Each canal is built by a single cell with intracellular lumen (sometimes canals are projections of the sinus cell, e.g. Nelson et al., 1983), which can show spatial differentiation into smooth and microvillar portions (McLaren, 1974; Waddell, 1968). Additional glandular, sphincter, or lip cells can be associated with the terminal duct (Chitwood and Chitwood, 1950; McLaren, 1974; Nelson et al., 1983; Waddell, 1968). The molecular phylogeny of nematodes shows that species with an H-system (or its modification) represent a monophyletic group whereas species with single cell excretory gland form a basal grade (Holterman et al., 2006; Smythe et al., 2019); hence, the latter should be considered as a plesiomorphic nematode arrangement, whereas the former represent synapomorphy of Secernentea. Both systems are nematode specific and represent derived excretory organs not easily comparable with any structures found in remaining Nephrozoa.

Interestingly, the putative unicellular secretory excretory organs were also described in the acoel *Paratomella rubra* (Ehlers, 1992), likely representing an evolutionary gain of excretory organs independent from nephrozoan lineage (Andrikou et al., 2019). The so-called dermonephridia are specialized epidermal cells, randomly distributed in the epidermis, which lack cilia but possess modified long microvilli on their apical surfaces and intracellular lacunar systems of tubules and vacuoles, which communicate with the exterior. Taking into account that dermonephridia are known only from a single acoel species, they probably represent a recent evolutionary novelty.

7.2.2 Protonephridia

Protonephridia use ultrafiltration through extracellular filters driven by ciliary action (Bartolomaeus and Ax, 1992; Ichimura and Sakai, 2017; Ruppert, 1994; Ruppert and Smith, 1988; Schmidt-Nielsen, 1997; Schmidt-Rhaesa, 2007; Figure 7.1). Some of the simplest protonephridia in the entire animal kingdom are found in Gnathostomulida (Figure 7.2; Lammert, 1985), a group of microscopic marine worms closely related to rotifers. Each of the gnathostomulid protonephridia consists of only three cells and can be used as an example of the minimal and most basic cellular architecture required for this type of organ (Figure 7.2; Lammert, 1985). The most proximal (or terminal) cell has a single cilium surrounded by the cytoplasmic portion (so-called filtering area) traversed by irregular clefts closed with a filtering membrane (i.e.

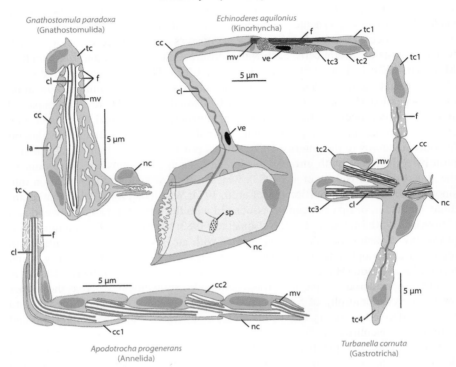

FIGURE 7.2 Morphological diversity of protonephridia. Schematic reconstruction of various protonephridial systems as inferred from TEM studies, based on Kristensen and Hay-Schmidt, 1989; Lammert, 1985; Teuchert, 1973 and Westheide, 1985. In all drawings, primary urine filtering cells (terminal cells) are in orange, primary urine modifying cells (canal cells) in blue, nephroporus cells in green, cell nuclei in dark grey, and intracellular cavities in light grey. Abbreviations: **cc** canal cell, **cl** cilium, **f** protonephridial filter, **la** lacunar system, **mv** microvilli, **nc** nephroporus cell, **sp** sieve plate (cuticular opening for protonephridium), **tc** terminal cell, **ve** vesicle. Number after abbreviation indicates multiplication of particular cell type.

slit diaphragm). The tubular filtering area tightly adjoins to the second cell of the system—the canal cell—that harbors the intracellular canal into which the cilium of the terminal cell protrudes. The canal cell is non-ciliated and has a complicated lacunar system and microvilli facing the canal lumen, which both greatly increase the cellular surface. Laterally, the canal cell adjoins the nephroporus cell, which is a modified epidermal cell, through which a canal of the protonephridium communicates with the exterior. The terminal cell provides the filter as well as the negative pressure necessary for the ultrafiltration and is therefore the site of primary urine production. The canal cell, on the other hand, is responsible for the secondary modification of the urine. Importantly, both components of the organ show ultrastructural

Cell Types, Morphology, and Evolution of Animal Excretory Organs

characteristics of epidermal cells, indicating that the entire organ has an ectodermal origin (Lammert, 1985).

There are two ways in which this extremely simple system can be complicated in other protonephridia-bearing animals. First of all, particular cells can be multiplied, resulting in many terminal cells opening to the single canal cell (e.g. in some gastrotrichs, Figure 7.2, Kieneke et al., 2008; Kieneke and Hochberg, 2012; Teuchert, 1973 or some kinorhynchs, Figure 7.2, Kristensen and Hay-Schmidt, 1989), one terminal cell can open to the canal built of many similar canal cells (e.g. in annelid *Apodotrocha*, Figure 7.2; Westheide, 1985), or both elements can be multiplied (e.g. in some Kinorhyncha and Loricifera; Neuhaus, 1988; Neuhaus and Kristensen, 2007, or in Nemertea, Bartolomaeus, 1985). When terminal cells are multiplied, each of them might form a separate filtering unit (Bartolomaeus, 1985; Teuchert, 1973), or they can be merged together, with the filtering area formed by two or more adjacent cells (Kieneke et al., 2008; Kieneke and Hochberg, 2012; Kristensen and Hay-Schmidt, 1989; Neuhaus, 1988; Neuhaus and Kristensen, 2007).

Apart from cell multiplication, additional cell types might be present in the protonephridial organs of some animals, increasing spatial diversification and functional specification of the nephridium. For instance, in some platyhelminths, the multicellular canal is divided into the proximal ciliated zone to which the terminal cells open and a distal zone, built exclusively by cells with extensive microvilli on the luminal surface (Rohde and Watson, 1993; Scimone et al., 2011; Vu et al., 2015). Additionally, it has been demonstrated, that in planarians, those two parts express different sets of solute carrier transporters and apparently have different roles in the secondary urine modification (Vu et al., 2015). In microscopic rotifers, the canal is built of two or three distinct cell types (Ahlrichs, 1993a, b; Warner, 1969), and additionally, in monogononts, it opens to the muscular bladder, which collects urine from paired protonephridial systems (Ahlrichs, 1993b; Warner, 1969). The complexity of the rotifer excretory system is further increased due to the fact that some of its parts are of a syncytial nature.

Protonephridia are present in many invertebrate taxa (sometimes only in larvae e.g. in phoronids, some molluscs, and annelids; see Baeumler et al., 2011; Bartolomaeus, 1989; Bartolomaeus and Quast, 2005; Goodrich, 1945; Hay-Schmidt, 1987; Koch et al., 2014; Ruthensteiner et al., 2001; Temereva and Malakhov, 2006; Todt and Wanninger, 2010), and even though they are often considered plesiomorphic (e.g. Bartolomaeus and Ax, 1992), they show remarkable diversity and evolutionary variation related to the fact that their particular elements can be easily organized in various ways (Ichimura and Sakai, 2017). For example, the filter might be built by a single cell with irregular openings, a single cell with slits, two adjacent terminal cells with interdigitating processes, or a terminal cell and a canal cell forming a common weir (Kieneke et al., 2008). Additionally, it might stand with or without a diaphragm or with or without supporting microvilli, and microvilli might be differentiated or uniform. The terminal cell might be monociliated (also called a solenocyte), multiciliated with independent cilia (then known as cyrtocyte), or multiciliated with all the cilia forming a single structure, the so-called flame as in rotifers or some platyhelminthes (Riemann and Ahlrichs, 2010; Rohde, 1991). The canal might be intracellular or multicellular, with or without cilia, spatially diversified or uniform. However, despite all their diversity, protonephridia are always straightforwardly comparable to each other (Bartolomaeus and Ax, 1992), and the primary homology statements regarding their particular portions can be readily made. The distribution of the protonephridia-bearing animals on the phylogenetic tree (Figure 7.1), suggests that those organs evolved either twice, once in Scalidophora and once in Spiralia, or are homologous between those two groups. This makes protonephridia an excellent model for studying the evolution of complexity and functionality of the excretory organs over long evolutionary time, combining morphological, cellular, and molecular levels.

7.2.3 Metanephridial Systems

The typical metanephridia (Figures 7.1 and 7.3) are present in some annelids (e.g. Bartolomaeus and Quast, 2005; Goodrich, 1945), in brachiopods (Kuzmina and Malakhov, 2015; Lüter, 1995), adult phoronids (Bartolomaeus, 1989; Storch and Herrmann, 1978; Temereva and Malakhov, 2006), cephalochordates (Ichimura and Sakai, 2017; Moller and Ellis, 1974; Ruppert, 1994), and adult hemichordates (Balser and Ruppert, 1990; Dilly et al., 1986; Mayer and Bartolomaeus, 2003). Furthermore, the onychophoran nephridia (Mayer, 2006), coxal glands of Chelicerata (Briggs and Moss, 1997; Koch et al., 2014), antennal gland of Crustacea (Bartolomaeus et al., 2009; Khodabandeh et al., 2005), molluscan heart-kidneys (Baeumler et al., 2011; Bartolomaeus, 1997), and the axial organ of echinoderms (Balser and Ruppert, 1993; Ezhova et al., 2013, 2014; Welsch and Rehkamper, 1987; Ziegler et al., 2009) can be also considered as metanephridial excretory systems, at least from a functional point of view. The ultrafiltration occurs in invertebrate metanephridial systems (Figure 7.3) between blood vessels (or their functional equivalents, e.g. haemocoelic sinus) and the coelomic lining where specialized cells—podocytes (Figure 7.3)—are present (Bartolomaeus and Ax, 1992; Ichimura and Sakai, 2017; Ruppert, 1994; Ruppert and Smith, 1988; Schmidt-Rhaesa, 2007). The primary urine is therefore synonymous with the coelomic fluid in those animals (Bartolomaeus and Ax, 1992; Ruppert, 1994; Schmidt-Rhaesa, 2007). The fluid leaves the coelom through the spatially independent structure, a metanephridium (Figure 7.3), which often consists of the proximal dilated portion (i.e. ciliated funnel) and the distal region (i.e. nephroduct), which might be further subdivided into regions with differing functions and cell type composition (Goodrich, 1945; Schmidt-Rhaesa, 2007). Therefore, compared to protonephridia, many different cell types build metanephridial systems, and their exact number and qualitative composition varies from taxon to taxon, showing diverse ultrastructural, developmental, and functional characteristics. In some animals—e.g. phoronids or some annelids—the metanephridia are ontogenetically predated by protonephridia (Bartolomaeus, 1989; Bartolomaeus and Quast, 2005; Goodrich, 1945; Koch et al., 2014; Temereva and Malakhov, 2006). In such instances, the terminal cells of the protonephridium degenerate during metamorphosis, and the adult metanephridium develops from the larval protonephridial canal, whereas the podocytes develop *de novo* from the myoepitheliocytes (Bartolomaeus, 1989; Bartolomaeus and

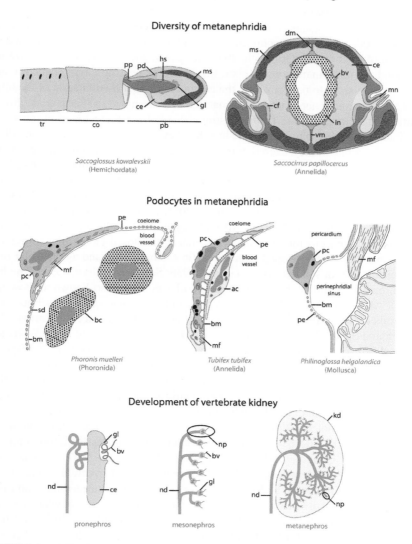

FIGURE 7.3 **Morphological diversity of metanephridial systems.** Schematic drawings of metanephridial systems in hemichordate (longitudinal section) and annelid (cross section), based on Balser and Ruppert, 1990 and Goodrich, 1945. Schematic drawings of podocytes present in various invertebrates reconstructed from TEM sectioning, based on Bartolomaeus, 1997; Peters, 1977; and Storch and Herrmann, 1978. Schematic organization of three developmental stages of vertebrate excretory organs adapted from Vize et al., 1997. In all drawings, blood vessels are in red, sites of primary urine filtration in orange, and sites of primary urine modification in blue. Abbreviations: **ac** amebocyte, **bc** blood cell, **bm** basal membrane, **bv** blood vessel, **ce** coelom, **cf** ciliated funnel, **co** collar region, **dm** dorsal mesentery, **gl** glomerulus, **hs** heart sinus, **in** intestine, **kd** kidney, **mf** myofibrils, **mn** metanephridium, **ms** musculature, **nd** nephric duct, **np** nephron, **pb** proboscis, **pc** podocyte, **pd** protocoel duct, **pe** pedicle, **pp** proboscis pore, **sd** slit diaphragm, **tr** trunk region, **vm** ventral mesentery.

Quast, 2005; Bartolomaeus et al., 2009; Storch and Herrmann, 1978; Temereva and Malakhov, 2006). The term nephromixium is sometimes used to refer to such a definite organ of dual origin (Goodrich, 1945; Temereva and Malakhov, 2006). On the other hand, in the Panarthropoda taxa, the podocyte-bearing cavity, called a sacculus, does not seem to be formed by the coelom but rather by dilatation of the blind end of the developing nephroduct (summarized and discussed in (Koch et al., 2014)). Nephridia in Arthropoda are further distinguishable from those found in other animals (including onychophorans) by the fact that they lack any ciliation (Mayer, 2006).

In the vertebrate kidney, excretion is carried out in a structural unit called a nephron (Figure 7.1; e.g. Gérard, 1936; Ichimura and Sakai, 2017; Ruppert, 1994; Vize et al., 1997), which, due to its importance in human physiology, is very well studied on morphological, physiological, developmental and molecular levels (e.g. Bates and Sims-Lucas 2016; Desgrange and Cereghini, 2015; Dressler, 2006; Lindström et al., 2018; Little et al., 2010; McMahon, 2016; Quaggin and Kreidberg, 2008). There are three types of nephrons found among vertebrates: the closed glomerular (present in the mammalian kidney), opened glomerular (in salamanders), and aglomerular ones (found exclusively in some teleosts) (Gérard, 1936; Schmidt-Nielsen, 1997). The first two are relatively similar: both have a glomerular portion (also known as Bowman capsule), where filtering cells—podocytes—tightly surround capillaries lined with the extremely thin-walled fenestrated endotheliocytes (Bates and Sims-Lucas 2016; Ichimura and Sakai, 2017; Koriyama et al., 1992; Schmidt-Nielsen, 1997; Wolff and Merker, 1966). The primary ultrafiltrate passes through the endothelium-ECM-podocytes barrier and is subsequently accumulated inside the capsule, from where it is drained by the proximal nephron tubule. Aside from the podocytes and endotheliocytes, two additional cell types of predominantly supportive function are found in the glomerulus—mesangial cells, which support capillaries (but also contribute to filtration of some molecules directly from the bloodstream) and parietal cells, which exclusively build the external wall of the Bowman capsule (Dressler, 2006; Ichimura and Sakai, 2017; Quaggin and Kreidberg, 2008; Vize et al., 1997). The nephron tubule leads from the glomerulus to the collective duct, which eventually opens to the urinary bladder. The tubule is differentiated into proximal, intermediate, and distal regions, which differ in function and, consequently, in the cell types of which they are composed (Bates and Sims-Lucas 2016; Desgrange and Cereghini, 2015; Lindström et al., 2018; Little et al., 2010; Schmidt-Nielsen, 1997). The main difference between closed and opened glomerular nephrons is that in the latter, the additional ciliated canal (composed of yet another cell type) connects its proximal tubule with the peritoneal cavity (Gérard, 1936). The aglomerular nephron, on the other hand, is found only in some, mostly marine, teleosts (Gérard, 1936; Ichimura and Sakai, 2017; Schmidt-Nielsen, 1997). It lacks a glomerular capsule and, in fact, uses only active transport through the cells for primary urine production (Bulger, 1965; Dobbs and Devries, 1975; Schmidt-Nielsen, 1997), which makes it actually an example of a secretory excretory organ. It is, however, homologous to the glomerular nephron, and, in some species, it develops ontogenetically from the glomerular condition (Gérard, 1936).

Depending on the arrangement of nephrons in the organ, it is possible to distinguish between a mesonephros and a metanephros (Figure 7.3; Bates, 2016; Ichimura and Sakai, 2017; Vize et al., 1997). In amniotes, they form an ontogenetic series with the metanephros being a definite excretory organ and the mesonephros found only in the embryonic or larval stages (Bates, 2016; Vize et al., 1997). The earliest developmental stage of the vertebrate kidney—the pronephros—does not have separate nephrons; the glomerulus and nephric duct are spatially separated, whereas the primary urine is accumulated inside the coelom (Figure 7.3; e.g. Møbjerg et al., 2000; Vize et al., 1997). From a functional perspective, it can therefore be categorized as the proper metanephridial system (Ichimura and Sakai, 2017).

The most distinctive cell type that seems to be shared by all metanephridial systems (including kidneys) is a podocyte (Ichimura and Sakai, 2017; Ruppert, 1994). Cells of this type are divided into the cell body and pedicles—long projections, which interdigitate with each other (Figure 7.3). On the junctions of the pedicles, the proteins anchored in the podocyte cell membrane might form the filtering membrane—a slit diaphragm (Figure 7.3; e.g. Gerke et al., 2003; Ichimura and Sakai, 2017; Quaggin and Kreidberg, 2008; Storch and Herrmann, 1978; Tryggvason and Wartiovaara, 2001). Interestingly, a filtering region of terminal protonephridial cells of planarians shares some ultrastructural and molecular similarities with the filtering portion of the podocyte (Ichimura and Sakai, 2017; Vu et al., 2015). Taking into account that metanephridia probably evolved independently at least few times in the animal kingdom (Figure 7.1; Bartolomaeus, 1997; Bartolomaeus and Ax, 1992; Koch et al., 2014), the podocytes of various animal groups represent analogous cell types (Bartolomaeus and Ax, 1992), which likely evolved by independent modification of the same ancestral filtering mechanism (e.g. from terminal cells of protonephridia; Ruppert, 1994).

7.2.4 Excretory Cells

In addition to multicellular excretory organs (and sometimes instead of them), some animals possess specialized cells, which serve excretory purposes (Haszprunar, 1996; Ruppert, 1994). Instead of expelling toxic waste products from the animal body, these cells accumulate the excreted substances inside their cytoplasm. Such accumulative excretory cells have been reported in a wide range of animals (Figure 7.1), and although they are generally referred to as nephrocytes, their ultrastructure and function differ from taxon to taxon.

The most studied excretory cells among nephrocytes are cells present in arthropods and onychophorans (Coons et al., 1990; Crossley, 1972; El-Shoura, 1989; Hessler and Elofsson, 1995; Seifert and Rosenberg, 1977; Shatrov, 1998; Vandenbulcke et al., 1998; Weavers et al., 2009; Miyaki et al., 2020). These cells, located inside haemolymph-filled cavities, are surrounded by ECM and resemble isolated podocytes with diaphragm-like structures. The waste products and other toxic compounds are filtered through the ECM and diaphragm and accumulate inside the nephrocytes (Vandenbulcke et al., 1998; Weavers et al., 2009). Insects have two sets of nephrocytes with similar ultrastructural properties—garland cells around the esophagus and pericardial cells located on the walls of the heart (Crossley, 1972; Weavers et al., 2009). Cells with similar ultrastructure and function have been also reported in mollusks, where they are called rhogocytes (Haszprunar, 1996; Rivest, 1992; Ruthensteiner et al., 2001). It has been proposed, based on molecular similarities (homology of the proteins forming the diaphragm) and intermediate morphological forms, that both arthropod nephrocytes and molluscan rhogocytes are actually homologues of the filtering nephridial cells, which became spatially separated from an ancestral excretory organ (Haszprunar, 1996; Hessler and Elofsson, 1995; Rivest, 1992; Ruppert, 1994; Weavers et al., 2009; Miyaki et al., 2020).

A very different cell type is the nephrocyte in tunicates. Tunicate nephrocytes represent a fraction of blood cells, which accumulate nitrogenous waste products inside voluminous vacuoles (Ballarin and Cima, 2005; Cima et al., 2014; George, 1939). A similar type of excretory cell is also present in the coelomic fluid of Bryozoa (Schwaha et al., 2020). Unlike the excretory cells of insects and mollusks, nephrocytes in tunicates and bryozoans lack both ECM and a slit diaphragm, and they represent specialized haemo- and coelomocytes; therefore, they are evolutionarily unrelated to other excretory organs.

7.2.5 Excretory Organs and Adaptations to New Environments

Excretory organs are not only responsible for the expulsion of metabolic waste products, but they are also the primary organs for osmoregulation and ion balance (Larsen et al., 2014; Schmidt-Nielsen, 1997; Schmidt-Rhaesa, 2007). The maintenance of water balance is especially challenging in environments such as freshwater and terrestrial habitats, where water is excessive or sparse, respectively (Schmidt-Nielsen, 1997). Another problem related to osmoregulation is faced by the organisms inhabiting the brackish and intertidal realms, where salinity can rapidly and dramatically change in a short time span, which requires the ability to accommodate to different salinity regimes (Schmidt-Nielsen, 1997). Moreover, differences in the environmental salinity and water availability result in adaptations regarding the mechanisms of ammonia (one of the most toxic metabolites) excretion, such as transforming it into less harmful nitrogenous end products (e.g. urea or uric acid) (Larsen et al., 2014; Needham, 1935; Schmidt-Nielsen, 1997; Weihrauch and O'Donnell, 2017). Therefore, modifications of excretory organs are believed to be crucial for colonizing new environments (Schmidt-Rhaesa, 2007). In fact, studies on various nephrozoans have shown that species inhabiting terrestrial, freshwater, brackish, and intertidal environments exhibit species-specific adaptations in the morphology (e.g. Krishnamoorthi, 1963; Randsø et al., 2019; Smith, 1984; von Nordheim and Schrader, 1994) and physiology (e.g. Generlich and Giere, 1996; Needham, 1935; Schmidt-Nielsen, 1997; Smith, 1970; Werntz, 1963) of their excretory organs, and these adaptations do not necessarily reflect their evolutionary relationship.

7.3 MOLECULAR IDENTITY OF EXCRETORY ORGANS: FROM DEVELOPMENT TO FUNCTION

In contrast to the large number of detailed morphological descriptions of excretory organs in a variety of animals, molecular data are scarce. Most of the gene expression studies have been conducted in vertebrates (mainly mammals, fish, and frogs) (summarized in Desgrange and Cereghini, 2015) and a handful of invertebrates (mainly planarians and flies) (Scimone et al., 2011; Vu et al., 2015; Weavers et al., 2009) and have revealed not only common transcriptional programs governing the development of excretory organs but also analogous sub-localization of solute transporters and structural proteins within the differentiated excretory compartments (Figure 7.4). Moreover, physiological studies in a large array of animals have shown that a number of ammonium transporters and proton pumps have conserved roles in excretory processes (Weihrauch and O'Donnell, 2017).

7.3.1 Nephron Development

Kidneys, the excretory organs of vertebrates, emerge from the intermediate mesoderm and develop through a sequential formation of up to three organs: pronephros, mesonephros, and metanephros (Figure 7.3; Saxen and Saxen, 1987). Pronephros and mesonephros are only transient structures in amniotes and function as the fetal excretory organs, whilst metanephros is the definitive adult kidney. In fish and frogs, the adult kidney is the mesonephros that replaces the embryonic pronephros. Both mesonephros and metanephros are composed from a basic structural and functional unit, the nephron, which shows a comparable regional organization (Wingert and Davidson, 2008). The formation of kidneys is dictated by similar genetic interactions and morphogenetic events (Desgrange and Cereghini, 2015) (Figure 7.4).

Nephron development in mammals starts with a mesenchyme-to-epithelial transition (MET) of the anterior intermediate mesoderm (IM) and the specification of renal progenitor cells. This process is induced by the interplay of an ectodermal BMP4 signaling (James and Schultheiss, 2005) and the expression of *odd skipped related gene (osr1)* (James et al., 2006) and *lhx1* transcription factors (Tsang et al., 2000). *Pax2* and *pax8* genes are activated shortly after and act redundantly in the nephric lineage specification (Bouchard et al., 2002). The renal progenitor cells will form an epithelial tubule, the future nephric duct. As the tubule extends posteriorly, a number of transcription factors, such as *hox11, six1/2/4, eya, sall, pax2, foxc*, and *wt1* (Brophy et al., 2001; Kobayashi et al., 2008; Kreidberg et al., 1993; Kume et al., 2000; Nishinakamura et al., 2001; Sajithlal et al., 2005; Wellik et al., 2002; Xu et al., 2003) are expressed along its anterior-posterior axis and induce the expression of *gdnf* (Gong et al., 2007; Moore et al., 1996). Ret/Gdnf signaling is crucial for the outgrowth and branching of the ureteric bud (UB) at the posterior end of the nephric duct, which grows into medially positioned metanephric mesenchyme, which in turn gives rise to a renal vesicle (Costantini and Shakya, 2006). Other signaling pathways involved in UB formation and branching include Wnt (Bridgewater et al., 2008; Carroll et al., 2005; Kispert et al., 1998), sonic hedgehog (Shh) (Cain and Rosenblum, 2011), bone morphogenic protein (Bmp) (Nishinakamura and Sakaguchi, 2014), and fibroblast growth factor (Fgf) (Bates, 2011). Eventually, each renal vesicle forms a nephron through a series of morphogenetic movements and patterning events. *Mafb, wt1*, and *lmx1b* drive podocyte specification (Miner et al., 2002; Moriguchi et al., 2006), and Notch signaling together with *pou3f3, hnf1b, irx1*, and *irx2* is responsible

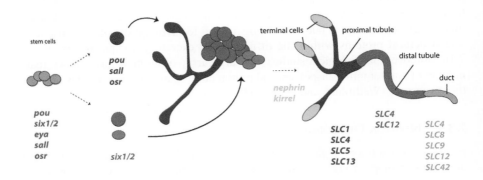
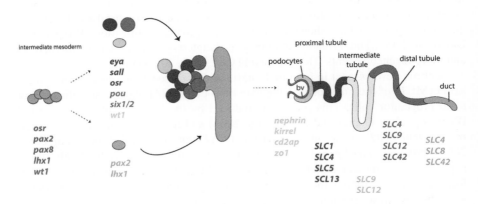

FIGURE 7.4 Development and structural correspondences of protonephridia and kidneys. Cartoon depiction of the molecular programs governing the regeneration of planarian protonephridia and the development of vertebrate kidneys, based on Scimone et al., 2011 and Vu et al., 2015. The corresponding structural components of protonephridia (terminal cell, tubule, duct) and kidneys (podocyte, tubule, duct) and the expression domains of orthologous genes in relation to their components, are color coded. Abbreviations: **bv** blood vessel.

for the proximal tubule fates (Cheng et al., 2007; Heliot et al., 2013; Nakai et al., 2003). The specification of the distal tubule is controlled from the molecular interplay of several transcription factors, including *lef1, sox9,* and *lhx1* (Mugford et al., 2009).

In the zebrafish *Danio rerio*, the renal progenitor field forms from the lateral-most IM that expresses the transcription factors *pax2a* and *pax8* (Pfeffer et al., 1998). *Osr1* has also a conserved expression since its endodermal expression during gastrulation promotes renal lineages at the expense of blood/vascular ones (Mudumana et al., 2008). The renal progenitors adopt an epithelial state through MET and form an epithelial tubule. Cells located in the anterior-most domain express *wt1a, wt1b, osr1, foxc1a,* and *lhx1a* (O'Brien et al., 2011; Perner et al., 2007; Tomar et al., 2014) and will form the podocytes, whilst the remaining cells will give rise to the proximal and distal tubule and express *jagged, irx3b, evi1,* and *pou3f3a/pou3f3b* (Li et al., 2014; Ma and Jiang, 2007; Wingert et al., 2007). The transcriptional interplay responsible for the formation of this boundary consists of *pax2a*, which forms a negative feedback loop with *wt1a* (Majumdar et al., 2000) and *hnf1b*, a suppressor of *pax2a* (Naylor et al., 2013). Tubule regionalization is also regulated by Retinoic Acid (Ra) signaling pathway (Wingert et al., 2007).

In the frog *Xenopus*, the renal progenitors emerge after a MET of the caudolateral IM resulting in the formation of a tubule. Once again, *osr1* and *osr2* are upstream of the specification of the renal progenitor field (Tena et al., 2007). The first renal molecular markers are *lhx1* and *pax8*, followed by the expression of *pax2, wt1,* and *hnf1β* (Buisson et al., 2015; Carroll and Vize, 1996, 1999; Carroll et al., 1999; Wild et al., 2000). *Wt1* specifies the future glomerulus whilst *pax2* is restricted to the future tubular region. The subdivision of the nephron into segments is governed by *evi1* expression in the distal tubule and pronephric duct (Van Campenhout et al., 2006) and *irx1, irx2,* and *irx3* expression in the proximal and intermediate tubule (Alarcon et al., 2008). The developing podocytes are specified from a cross talk between the *wt1, foxc2, lmx1b* and *mafb* genes (Haldin et al., 2008; White et al., 2010). Signaling pathways with crucial roles in *Xenopus* nephrogenesis involve Wnt (Lavery et al., 2008; Saulnier et al., 2002; Tetelin and Jones, 2010), Fgf (Urban et al., 2006), Bmp (Bracken et al., 2008), Notch (McLaughlin et al., 2000; White et al., 2010), and Ra (Cartry et al., 2006).

7.3.2 Protonephridial Development

Although there are several morphological descriptions on protonephridial development (Baeumler et al., 2011; Bartolomaeus, 1985; Hasse et al., 2010; Rohde et al., 1988; Temereva and Malakhov, 2006; Wenning et al., 1993), molecular studies are extremely limited. The most detailed work has been performed on the planarian *Schmidtea mediterranea* (Scimone et al., 2011) that has shown a remarkable conservation of the molecular programs between the regeneration of planarian protonephridia and the development of vertebrate kidneys (Figure 7.4). Planarian protonephridia consist of four cells types, the flame (terminal) cell, the ciliated tubule cell type (proximal tubule), the tubule-associated cell type (distal tubule), and the duct. Regeneration and RNAi experiments on amputated animals showed a conserved function of *eya, six1/2, pou3, hunchback, sall,* and *osr* genes in the regeneration and maintenance of protonephridia. *Eya, six1/2, pou3, sall,* and *osr* are expressed in the specified progenitor cells, whose fate segregates and results in the formation of the ciliated tubule cell type and the tubule-associated cell type. The ciliated tubule cell type continues to express *pou3, sall,* and *osr*, whilst the tubule-associated cells express *six1/2* (Scimone et al., 2011).

The conserved role of *pou3* is also reported in the nematode *C. elegans*, where an orthologous gene, *ceh-6*, is required for the formation and function of its excretory cell (Burglin and Ruvkun, 2001). Other conserved nephrogenesis-related transcription factors have been shown to be *pax3* in the developing nephridia of the leech *Helobdella robusta* (Woodruff et al., 2007), *pax2/5/8* in the nephridium of the cephalochordate *Branchiostoma floridae* (Kozmik et al., 1999), and *sall* in the protonephridium precursors of the polychaete *Hydroides elegans* (Arenas-Mena, 2013).

7.3.3 Development of Insect Malpighian Tubules

The Malpighian tubules of insects consist of two cell types: the primary cell (PC) and the secondary (stellate) cell (SC). In *Drosophila*, these two cell types have a different developmental origin; while the PC derives from an ectodermal primordium at the hindgut/midgut junction and forms the Malpighian tubule epithelium, the SC originates from the posterior mesoderm and invades the tubule epithelium through MET (Denholm et al., 2003). The molecular patterning of Malpighian tubule development appears quite different compared to the protonephridia and kidneys. *Krüppel* and *cut* expression in a cluster of cells of the hindgut mark the onset of PC development due to Wnt signaling. Once specified, this cluster of cells starts to form bud-like branches under the control of Decapentaplegic (BMP) signaling and Brinker (Hatton-Ellis et al., 2007). Later on, a regulatory interplay between *cut, barr, three rows, pebble, pimples, trachealess, ribbon, raw, crooked neck, faint sausage, pant*, and *schnurri*, under a DPP/BMP signal, will result in the formation and morphogenesis of tubules (Hatton-Ellis et al., 2007; Jack and Myette, 1999; Shim et al., 2001). As the tubules elongate, a caudal mesodermal population that will become SC expresses the transcription factors *tiptop* and *teashirt*, undergoes MET, and intercalates in the tubule (Denholm et al., 2003). The developmental process of Malpighian tubule development is overall conserved, since studies in *Tribolium* have shown that the onset, morphogenesis, and molecular profile are similar. An important difference is the fact that in *Drosophila*, only differentiated PC express *cut*, whilst in *Tribolium cut* is expressed in both PC and SC (King and Denholm, 2014).

7.3.4 Molecular Basis of Functional Compartmentalization

The terminal differentiated excretory organs are highly specialized and compartmentalized into discrete segments, composed of distinct epithelial cell types that carry out different functions. Each region is characterized by the expression of a set of structural proteins and transporters, involved in the ultrafiltration and modification of excreted material, such as glucose and solute transport, and homeostasis. The spatial distribution of these proteins on the different segments of excretory organs is, once more, remarkably conserved (Figure 7.4; Kozmik et al., 1999; Vu et al., 2015; Weavers et al., 2009).

In vertebrates, the nephron is divided into five segments: the glomerulus, the proximal, intermediate, and distal tubule, and the collecting duct. The glomerulus has a central role in ultrafiltration and expresses a set of membrane-associated proteins, such as Nephrin (NPHS1), KIRREL1 (NEPH1), CD2AP, ZO1, Nck, and Stomatin/Podocin (summarised in Patari-Sampo et al., 2006). The tubular segments are further subdivided into smaller regions, each of them specialized in different aspects of modifying the excreted fluid. The segmental organization is highly conserved, with the solute carrier (SLC) protein family sequentially expressed along the tubular segments of different types of nephrons, defining their boundaries (Desgrange and Cereghini, 2015). SLCs are membrane transporters, composed of 52 families, which can transport a number of different substrates (Hediger et al., 2013). In both the mammalian metanephros and *Xenopus* pronephros, subsets of cells of the proximal tubule are specialized in reabsorbing: a) salts, expressing members from the bicarbonate transporter SLC4, sodium- and chloride-dependent neurotransmitter transporter SLC6, sodium/proton exchanger SLC9, sodium-sulfate/carboxylate cotransporter SLC13, and organic cation/anion/zwitterion transporter SLC22 families; b) amino acids, expressing members from the glutamate and neutral amino acid transporter SLC1, heavy subunits of the heteromeric amino acid transporter SLC3, and cationic/glycoprotein-associated amino acid transporter SLC7 families; and c) glucose, expressing members from the sodium-glucose cotransporter SLC5 family (Landowski et al., 2008; Raciti et al., 2008). SLC9a3 expression also marks the proximal tubule of the pronephros of *D. rerio*, suggesting a similar function of this segment in reabsorbing sodium (Wingert et al., 2007). The intermediate tubule of the nephron (Henle's loop) modifies the concentration of excreted fluid and does not share an extensive molecular conservation among the different nephron types (Desgrange and Cereghini, 2015). Cells of the distal tubule are specialized in the reabsorption and secretion of ions, ammonium, and water through the SLC9 and the electroneutral cation-chloride cotransporter SLC12. Common transporters characterizing the distal tubule of the mammalian metanephros and *Xenopus* pronephros are members of SLC12, monocarboxylate transporter SLC16, type III sodium-phosphate cotransporter SLC20, and ammonium transporter SLC42/Rhesus families (Landowski et al., 2008; Raciti et al., 2008). Cells comprising the distal tubule of *Xenopus* also express SLC4a4 (Raciti et al., 2008), whilst in *D. rerio*, only members of the SLC12 family are seen (Wingert et al., 2007). Finally, the collecting duct consists of highly specialized, electrically tight cell types with dedicated roles in reabsorption and secretion of salts and water and the concentration and preparation of the urine. In mammals, the collecting duct expresses members of the sodium/potassium exchanger SLC8, SLC16, SLC20, SLC22, and SLC42/Rhesus families (Landowski et al., 2008), while in *Xenopus*, the analogous segment, the collecting tubule, also expresses SLC12a3 and not SLC42/Rhesus (Raciti et al., 2008). Even greater differences are seen in *D. rerio*, where the pronephric duct does not express any of these transporters (Wingert et al., 2007). The fact that the distal tubule of *D. rerio* and *Xenopus* pronephros expresses SLC12a3 and *Xenopus* does not express SLC42/Rhesus in their collecting tubule but only in their distal tubule, suggests that the function of the distal tubule of *D. rerio* and *Xenopus* might be analogous to a mammalian tubule/duct hybrid (Desgrange and Cereghini, 2015). Another family of transporters that exhibit a nephron segment-specific expression is the one of aquaporins, which transports mainly water, urea, and glycerol (Gomes et al., 2009). In mammals, the proximal tubule expresses the water/nitrate/chloride transporter AQ1 (Nielsen et al., 1993), the aquaglyceroporin AQ7 (Nejsum et al., 2000), the water/ammonia transporter AQ8 (Elkjaer et al., 2001), and the super-aquaporin AQ11 (Morishita et al., 2005), while the collecting duct expresses the water transporters AQ2 (Fushimi et al., 1993) and AQ4 (Terris et al., 1995), the aquaglyceroporin AQ3 (Ecelbarger et al., 1995), the water/nitrate/chloride transporter AQ6 (Yasui et al., 1999), and AQ8. In *Xenopus*, a similar distribution of the AQ2 and AQ3 mammalian orthologs has been shown in the collecting duct. However, AQ1 expression was restricted to the glomerulus, in contrast to what is observed in mammals (Pandey et al., 2010). The situation looks even more different in teleosts, where the proximal tubules express one AQ8, one AQ10-like, and one AQ3 paralog (Engelund and Madsen, 2015; Santos et al., 2004). Moreover, two copies of AQ1 have a renal expression but have acquired different functions: AQ1a is expressed in the proximal tubule and AQ1b in the distal tubule (Engelund and Madsen, 2015; Madsen et al., 2011). Finally, one AQ11 and one AQ12 ortholog are also expressed in the teleost nephron, but the exact expression domain has not yet been revealed (Kim et al., 2014; Madsen et al., 2014).

In the planarian *S. mediterranea*, protonephridia are divided into four major compartments: the flame (terminal) cells, the proximal tubule, the distal tubule, and the duct. The flame cells carry out ultrafiltration and express Nephrin/Kirrel. In a similar fashion to the metanephridial systems, the tubular compartments are further subdivided in smaller specialized domains with different cell types being defined by the expression of a suite of SLC transporters and having diverse roles in modifying excretion (Vu et al., 2015). For instance, the proximal tubule, primarily responsible for the recovery of filtered substances and reabsorption of salts, expresses members of the SLC1, SLC5, SLC4, SLC6, SLC13, and SLC22 families. The distal tubule has a central role in homeostasis and expresses members of the SLC4 and SLC12 families, which also mark the duct. The duct expresses SLC42/Rhesus as seen in mammals but not in *D. rerio* and *Xenopus*, as well as a number of other SLCs, which mark the proximal tubule of vertebrates, such as SLC6, SLC7, and SLC9, indicating a role of the planarian duct not only in the urine concentration but also in reabsorption of salts and amino acids. Interestingly, aquaporins are not expressed in any of the protonephridial compartments, suggesting divergent functions of these transporters in planarians (Vu et al., 2015).

Malpighian tubules of some insects (e.g. the fly *Drosophila melanogaster* and the mosquito *Aedes aegypti*) are also divided in distinct compartments with different physiological functions: the initial, transitional, main, and lower segments, and the ureter (Beyenbach et al., 1993; Sozen et al., 1997). Each segment is populated by a different number and positional combination of PC and SC cells, depending on the species investigated. Overall, the initial and transitional segments do not participate in secretion but rather act as storage segments and transport (Dow et al., 1994; Dube et al., 2000). The main segment has both secretory and absorptive roles for salts and water and generates the primary urine. The PC cells of the main segment express a basolateral sodium/potassium transporter (Torrie et al., 2004), an inward-rectifier potassium transporter (Kir) (Evans et al., 2005), sodium/proton exchangers of the NHA and NHE (SLC9) families (Pullikuth et al., 2006; Rheault et al., 2007),

Cell Types, Morphology, and Evolution of Animal Excretory Organs

and an aquaporin (Kaufmann et al., 2005). In contrast, the SC cells, which mostly have an absorptive role, express mainly a chloride transporter and an aquaporin (Kaufmann et al., 2005; Kolosov and O'Donnell, 2020; O'Connor and Beyenbach, 2001; O'Donnell et al., 1998). The expression of an SLC4 exchanger has also been reported in the SC cells of some species (Linser et al., 2012; Piermarini et al., 2010). Finally, the lower segment and rectum are mostly dedicated in the reabsorption of salts and water and the concentration of urine (O'Donnell and Maddrell, 1995).

Molecular similarities found in the building blocks of the ultrafiltration sites of proto- and metanephridial systems are also seen in individual ultrafiltration cells. The membrane-associated protein complex composed of Nephrin, Kirrel, Cd2ap, Zo1, and Stomatin/Podocin also forms the slit diaphragm of nephrocytes of the fly *Drosophila melanogaster* (Weavers et al., 2009; Zhuang et al., 2009) that filters the haemolymph (Weavers et al., 2009; Wigglesworth, 1943; Zhang et al., 2013). Nephrocytes are not only involved in ultrafiltration but also in protein reabsorption, via receptors such as Cubilin and Amnionless (AMN), similar to mechanisms encountered in the renal proximal tubule cells (Zhang et al., 2013). Moreover, Nephrin/Kirrel expression has been described in the ultrafiltration apparatus of rhogocytes of the snail *Biomphalaria glabrata* (Kokkinopoulou et al., 2014), through which proteins and ions are filtered (Kokkinopoulou et al., 2015).

7.3.5 Nitrogenous Waste Excretion

One of the main and most toxic products of excretion in animals is nitrogenous waste. Nitrogenous waste products are the result of the amino acid catabolism (Campbell, 1991) and exist in three forms: ammonia, urea, or uric acid. Most aquatic species excrete ammonia, due to its high solubility in water, whilst semi-aquatic and terrestrial species usually convert ammonia to less hazardous and less soluble forms, such as urea and uric acid. Ammonia can occur either in its gaseous (NH_3) or ionic form (NH_4^+); however, due to the high pK of NH_3 in physiological solutions, the vast majority of ammonia exists as NH_4^+. The excretory process of ammonia has been investigated in a wide array of animals and shows extensive conservation of the repertoire of the ammonia transporters and proton pumps, independently of the presence of specialized excretory organs (Weihrauch and O'Donnell, 2017).

In kidneys, the epithelial cells of the proximal tubule secrete ammonia apically into the luminal fluid. A significant fraction of this ammonia secretion occurs via Na^+/NH_4^+ exchange by the sodium/proton exchangers SLC9 (NHE), and some also takes place via diffusion (Bourgeois et al., 2010; Knepper et al., 1989; Preisig and Alpern, 1990). Almost all secreted ammonia is reabsorbed by cells of the thick ascending limb of Henle's loop into the interstitium, both by K^+/NH_4^+ exchange of the apical cation-chloride cotransporter SLC12 (NKCC) (Good, 1994) and Na^+/NH_4^+ exchange of the basolateral SLC9 exchangers (Blanchard et al., 1998). Another cotransporter that plays an important role in NH_4^+ reabsorption is the basolateral bicarbonate transporter SLC4, which drives NH_3 diffusion across the basolateral membrane due to the bicarbonate transport into the cell and the subsequent rise of the intracellular pH (Good et al., 1984; Lee et al., 2010). The accumulated ammonia in the interstitium forms a gradient that drives diffusion across the epithelium of the cells of the collecting duct and secretion of the concentrated ammonia to the lumen. This process is also supported by a proton gradient formed by the apical vacuolar H$^+$-ATPase, which creates an acidic environment facilitating NH$_4^+$ entrapment (Flessner et al., 1991; Star et al., 1987) and a basolateral Na$^+$/K$^+$ ATPase (NKA) that actively transfers ammonia by Na$^+$/NH$_4^+$ exchange (Wall, 1996). Finally, the Rhesus ammonia transporters (RhBG and RhCG), spatially restricted to the collecting duct, also participate in secretion into the lumen and preparation of the urine (Mak et al., 2006).

Species-specific excretory organs/sites have also recruited these transporters for excreting ammonia. In Malpighian systems, ammonia is firstly secreted into the Malpighian tubules and then actively absorbed by the hindgut and midgut, which express the ammonia transporter Rhesus, sodium/proton exchangers (NHE), and a vacuolar H$^+$-ATPase (Blaesse et al., 2010; Weihrauch, 2006). Similarly, in the posterior rectum (anal papillae) of the aquatic mosquito larvae *Aedes aegypti*, the ammonia transporters Rhesus and the Rhesus-related AMTs, as well as a basal NKA and an apical vacuolar H$^+$-ATPase are involved in ammonia excretion (Chasiotis et al., 2016; Durant and Donini, 2018). In the excretory H-system of the nematode *C. elegans*, ammonia enters the excretory cells via the basolateral NKA and potassium transporters and then diffuses across the apical membrane through an acid-trapping mechanism (Adlimoghaddam et al., 2015). Rhesus and the vacuolar H$^+$-ATPase are likely also involved in ammonia excretion in the plicate organ of mussels of the Mytilidae family (Thomsen et al., 2016). Other animal-specific ammonia excretory sites include the gills of several crustacean species and the branchial appendages of the marine annelid *Eurythoe complanata*, which all seem to express NKA, the vacuolar H$^+$-ATPase, Rhesus, as well as AMTs (in the case of the *E. complanata*) (Si et al., 2018; Thiel et al., 2017; Weihrauch et al., 2017). Interestingly, NKA, NHE, Rhesus, and the vacuolar H$^+$-ATPase are also reported to be involved in ammonia excretion even when this process occurs through the integument, as shown in the leech *Nephelopsis obscura* (Quijada-Rodriguez et al., 2015), the planarian *S. mediterranea* (Weihrauch et al., 2012), and the nematode *C. elegans* (Adlimoghaddam et al., 2016), or through digestive-associated tissues, as suggested in members of Xenacoelomorpha and Cnidaria (Andrikou et al., 2019).

The astonishing molecular similarities reported, not only at the developmental level but also at the structural and functional level of excretory systems, has led some authors to propose their common evolutionary origin (Haszprunar, 1996; Ruppert, 1994; Scimone et al., 2011; Vu et al., 2015; Weavers et al., 2009). However, given the fact that correspondences in molecular patterning cannot be the only criterion for supporting homology, especially in the level of a complex organ system, such as a nephridium, these interpretations need to be handled with caution.

7.4 THE PROBLEM OF HOMOLOGY

Although excretory organs are present across the Nephrozoa, their emergence and evolutionary relationship remain unsolved. When mapping nephridia on the recent animal phylogeny, it is likely that the protonephridia are the ancestral form, from which at least the metanephridia evolved (Bartolomaeus and Ax, 1992; Ruppert, 1994). The evolution of metanephridia has likely happened multiple times independently, given that coeloms emerged convergently in different animal lineages (Koch et al., 2014). The fact that protonephridia can directly develop into metanephridia in some lineages indicates a close evolutionary relationship between these two types of excretory organs (Bartolomaeus and Ax, 1992; Ruppert, 1994). The homology of protonephridia and metanephridia is partly supported by similar transcription factors that seem to be involved in patterning both structures (Scimone et al., 2011). However, the taxon sampling so far is very narrow and therefore needs to be extended to solidify this interpretation.

So far, the cell type perspective does not contribute much to the question of the homology of excretory organs, mainly because it deals with a different level of homology (Abouheif, 1997). Firstly, excretory organs are composed of different cells that perform diverse functions. At what time in evolutionary history these cells originated and, if homologous, have been assembled to a functional organ, remains unclear. Secondly, the different cells that build a nephridium can perform similar functions with very different structures, which could speak for their convergence. When comparing, for example, the terminal ultrafiltration cell with other cell types that possess a "collar" or slit-like openings, it is evident that these collar-like structures within nephridia are highly diverse and therefore difficult to compare between species (Figure 7.5). Sometimes the filters are formed between the cells, sometimes the openings are within the cells, and in other cases, these slits are formed by

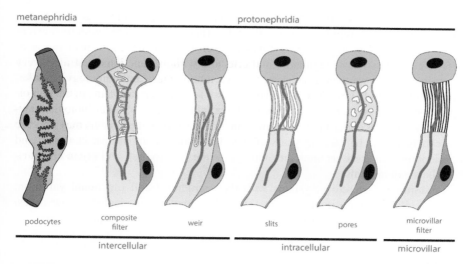

FIGURE 7.5 Diversity of filter-forming cells within nephridia. The filter apparatus can be formed by different cell features. Intercalated cells form the filtration apparatus between cells, filtration can be performed by cellular pores, and slits or microvilli can build up the filtration apparatus. Cells forming extracellular filters are depicted in orange, while nephroduct cells are in blue.

microvilli. All variations perform a similar task, namely the filtering of the primary urine. Do these differences suggest a convergence of the cells or even of the whole organ? On which level can these cellular structures be homologous? Can one use the presence of microvilli to state cell type homology, or can we only homologize the microvilli themselves? It is clear that a cell is composed out of many different substructures that might have to be treated separately when investigating their origin (Carvalho-Santos et al., 2011; Sebé-Pedrós et al., 2013). This raises the question of how many substructures are necessary to characterize a "cell type" and whether these are sufficient to homologize these cell types between species.

Additional problems are introduced when transcriptomic similarities are used for characterizing a cell type. The transcriptomic differences between cell-types and cell-states is a continuum, meaning that clear boundaries are established by the observer and are therefore artificial (Trapnell, 2015). It is furthermore in the nature of a cell that transcriptomic noise (and in some cases technical noise), which can be to some extent stochastic, obscures the potential signal that could be used to characterize the cell type/state (Ballouz et al., 2019). But what kind of signal are we looking for? As mentioned already, subcellular structures that are plesiomorphic for a clade (e.g. cilia, microvilli for Metazoa) may provide a signal but cannot be used to homologize cell-types. Some authors propose the use of a combination of transcription factors and effector genes to detect cell types within transcriptomes (Arendt et al., 2016). Considering what we know about gene regulatory networks and their flexibility and evolutionary exchangeability of key-regulators within these networks, using "this gene combination" as definitions of cell type can lead to wrong conclusions, especially when taking into account false positives and false negatives. Moreover, the concentration of transcription factors on the protein level may impact the output of a gene regulatory network and cannot yet be detected with the current single-cell methods (Marx, 2019). Finally, given the fact that the consideration of sets of co-expression of regulators and effectors without functional testing is very arbitrary, how do we then discriminate between homoplasy and homology? (Shafer, 2019; Tschopp and Tabin, 2017). In principle, we face similar problems in the homologization of cell types that we face with the homologization of other biological levels. Noise, drift of underlying structures, and methodological problems may obscure the conclusions. The coming years of data harvesting, comparative analyses, and developments in these technologies will guide the way for cell and organ comparisons between animals.

In summary, excretory systems with their structural and functional variation, diverse cellular composition, and variable embryology are an ideal showcase to test different approaches currently used for unravelling the origin of organ systems.

REFERENCES

Abouheif, E., 1997. Developmental genetics and homology: A hierarchical approach. *Trends Ecol Evol* 12, 405–408.

Adlimoghaddam, A., Boeckstaens, M., Marini, A.M., Treberg, J.R., Brassinga, A.K., Weihrauch, D., 2015. Ammonia excretion in *Caenorhabditis elegans*: Mechanism

and evidence of ammonia transport of the Rhesus protein CeRhr-1. *J Exp Biol* 218, 675–683.
Adlimoghaddam, A., O'Donnell, M.J., Kormish, J., Banh, S., Treberg, J.R., Merz, D., Weihrauch, D., 2016. Ammonia excretion in *Caenorhabditis elegans*: Physiological and molecular characterization of the rhr-2 knock-out mutant. *Comp Biochem Physiol A Mol Integr Physiol* 195, 46–54.
Ahlrichs, W., 1993a. Ultrastructure of the protonephridia of *Seison annulatus* (Rotifera). *Zoomorphology* 113, 245–251.
Ahlrichs, W.H., 1993b. On the protonephridial system of the brackish water rotifer *Proales reinhardti* (Rotifera, Monogononta). *Microfauna Marina* 8, 39–53.
Alarcon, P., Rodriguez-Seguel, E., Fernandez-Gonzalez, A., Rubio, R., Gomez-Skarmeta, J.L., 2008. A dual requirement for iroquois genes during *Xenopus* kidney development. *Development* 135, 3197–3207.
Andrikou, C., Thiel, D., Ruiz-Santiesteban, J.A., Hejnol, A., 2019. Active mode of excretion across digestive tissues predates the origin of excretory organs. *PLoS Biol* 17, e3000408.
Arab, A., Caetano, F.H., 2002. Segmental specializations in the Malpighian tubules of the fire ant *Solenopsis saevissima* Forel 1904 (Myrmicinae): An electron microscopical study. *Arthropod Struct Dev* 30, 281–292.
Arenas-Mena, C., 2013. Brachyury, Tbx2/3 and sall expression during embryogenesis of the indirectly developing polychaete *Hydroides elegans*. *Int J Dev Biol* 57, 73–83.
Arendt, D., Musser, J.M., Baker, C.V.H., Bergman, A., Cepko, C., Erwin, D.H., Pavlicev, M., Schlosser, G., Widder, S., Laubichler, M.D., Wagner, G.P., 2016. The origin and evolution of cell types. *Nat Rev Genet* 17, 744–757.
Baeumler, N., Haszprunar, G., Ruthensteiner, B., 2011. Development of the excretory system in the polyplacophoran mollusc, *Lepidochitona corrugata*: The protonephridium. *J Morphol* 272, 972–986.
Ballarin, L., Cima, F., 2005. Cytochemical properties of *Botryllus schlosseri* haemocytes: Indications for morpho-functional characterisation. *Eur J Histochem* 49, 255–264.
Ballouz, S., Pena, M.T., Knight, F.M., Adams, L.B., Gillis, J.A., 2019. The transcriptional legacy of developmental stochasticity. *bioRxiv*, 2019.2012.2011.873265.
Balser, E.J., Ruppert, E.E., 1990. Structure, ultrastructure, and function of the preoral heart kidney in *Saccoglossus kowalevskii* (hemichordata, enteropneusta) including new data on the stomochord. *Acta Zool* 71, 235–249.
Balser, E.J., Ruppert, E.E., 1993. Ultrastructure of axial vascular and celomic organs in comasterid featherstars (Echinodermata, Crinoidea). *Acta Zool* 74, 87–101.
Bartolomaeus, T., 1985. Ultrastructure and development of the protonephridia of *Lineus viridis* (Nemertini). *Microfauna Marina* 2, 61–83.
Bartolomaeus, T., 1989. Ultrastructure and relationship between protonephridia and metanephridia in *Phoronis muelleri* (Phoronida). *Zoomorphology (Berlin)* 109, 113–122.
Bartolomaeus, T., 1997. Ultrastructure of the renopericardial complex of the interstitial gastropod *Philinoglossa helgolandica* Hertling, 1932 (Mollusca: Opisthobranchia). *Zool Anz* 235, 165–176.
Bartolomaeus, T., Ax, P., 1992. Protonephridia and metanephridia – Their relation within the Bilateria. *Z Zool Syst Evolutionsforsch* 30, 21–45.
Bartolomaeus, T., Quast, B., 2005. Structure and development of nephridia in Annelida and related taxa. *Hydrobiologia* 535–536, 139–165.
Bartolomaeus, T., Quast, B., Koch, M., 2009. Nephridial development and body cavity formation in *Artemia salina* (Crustacea: Branchiopoda): No evidence for any transitory coelom. *Zoomorphology (Berlin)* 128, 247–262.

Bates, C.H.J., Sims-Lucas, S., 2016. Embryonic development of the Kidney. In: Al, A.E. (Ed.), *Pediatric Nephrology*. Berlin Heidelberg: Springer-Verlag, pp. 3–36.

Bates, C.M., 2011. Role of fibroblast growth factor receptor signaling in kidney development. *Pediatr Nephrol* 26, 1373–1379.

Berridge, M.J., Oschman, J.L., 1969. A structural basis for fluid secretion by malpighian tubules. *Tissue Cell* 1, 247–272.

Beyenbach, K., Oviedo, A., Aneshansley, D., 1993. Malpighian tubules of *Aedes aegypti*: Five tubules, one function. *J Insect Physiol* 39, 639–648.

Bitsch, C., Bitsch, J., 2004. Phylogenetic relationships of basal hexapods among the mandibulate arthropods: A cladistic analysis based on comparative morphological characters. *Zool Scripta* 33, 511–550.

Blaesse, A.K., Broehan, G., Meyer, H., Merzendorfer, H., Weihrauch, D., 2010. Ammonia uptake in *Manduca sexta* midgut is mediated by an amiloride sensitive cation/proton exchanger: Transport studies and mRNA expression analysis of NHE7, 9, NHE8, and V-ATPase (subunit D). *Comp Biochem Physiol A Mol Integr Physiol* 157, 364–376.

Blanchard, A., Eladari, D., Leviel, F., Tsimaratos, M., Paillard, M., Podevin, R.A., 1998. NH_4^+ as a substrate for apical and basolateral $Na(^+)$-H^+ exchangers of thick ascending limbs of rat kidney: evidence from isolated membranes. *J Physiol* 506(Pt 3), 689–698.

Bouchard, M., Souabni, A., Mandler, M., Neubuser, A., Busslinger, M., 2002. Nephric lineage specification by Pax2 and Pax8. *Genes Dev* 16, 2958–2970.

Bourgeois, S., Meer, L.V., Wootla, B., Bloch-Faure, M., Chambrey, R., Shull, G.E., Gawenis, L.R., Houillier, P., 2010. NHE4 is critical for the renal handling of ammonia in rodents. *J Clin Invest* 120, 1895–1904.

Bracken, C.M., Mizeracka, K., McLaughlin, K.A., 2008. Patterning the embryonic kidney: BMP signaling mediates the differentiation of the pronephric tubules and duct in *Xenopus laevis*. *Dev Dyn* 237, 132–144.

Bradley, T.J., 1983. Functional design of microvilli in the Malpighian tubules of the insect *Rhodnius prolixus*. *J Cell Sci* 60, 117–135.

Bridgewater, D., Cox, B., Cain, J., Lau, A., Athaide, V., Gill, P.S., Kuure, S., Sainio, K., Rosenblum, N.D., 2008. Canonical WNT/beta-catenin signaling is required for ureteric branching. *Dev Biol* 317, 83–94.

Briggs, R.T., Moss, B.L., 1997. Ultrastructure of the coxal gland of the horseshoe crab *Limulus polyphemus*: Evidence for ultrafiltration and osmoregulation. *J Morphol* 234, 233–252.

Brophy, P.D., Ostrom, L., Lang, K.M., Dressler, G.R., 2001. Regulation of ureteric bud outgrowth by Pax2-dependent activation of the glial derived neurotrophic factor gene. *Development* 128, 4747–4756.

Buisson, I., Le Bouffant, R., Futel, M., Riou, J.F., Umbhauer, M., 2015. Pax8 and Pax2 are specifically required at different steps of *Xenopus* pronephros development. *Dev Biol* 397, 175–190.

Bulger, R.E., 1965. The fine structure of the aglomerular nephron of the toadfish, *Opsanus tau*. *Am J Anat* 117, 171–191.

Burglin, T.R., Ruvkun, G., 2001. Regulation of ectodermal and excretory function by the *C. elegans* POU homeobox gene ceh-6. *Development* 128, 779–790.

Cain, J.E., Rosenblum, N.D., 2011. Control of mammalian kidney development by the Hedgehog signaling pathway. *Pediatr Nephrol* 26, 1365–1371.

Campbell, J., 1991. Excretory nitrogen metabolism. *Environ Metab Anim Physiol* 1, 277–325.

Carroll, T.J., Park, J.S., Hayashi, S., Majumdar, A., McMahon, A.P., 2005. Wnt9b plays a central role in the regulation of mesenchymal to epithelial transitions underlying organogenesis of the mammalian urogenital system. *Dev Cell* 9, 283–292.

Carroll, T.J., Vize, P.D., 1996. Wilms' tumor suppressor gene is involved in the development of disparate kidney forms: Evidence from expression in the *Xenopus* pronephros. *Dev Dyn* 206, 131–138.

Carroll, T.J., Vize, P.D., 1999. Synergism between Pax-8 and lim-1 in embryonic kidney development. *Dev Biol* 214, 46–59.

Carroll, T.J., Wallingford, J.B., Vize, P.D., 1999. Dynamic patterns of gene expression in the developing pronephros of *Xenopus laevis*. *Dev Genet* 24, 199–207.

Cartry, J., Nichane, M., Ribes, V., Colas, A., Riou, J.F., Pieler, T., Dolle, P., Bellefroid, E.J., Umbhauer, M., 2006. Retinoic acid signalling is required for specification of pronephric cell fate. *Dev Biol* 299, 35–51.

Carvalho-Santos, Z., Azimzadeh, J., Pereira-Leal, J.B., Bettencourt-Dias, M., 2011. Evolution: Tracing the origins of centrioles, cilia, and flagella. *J Cell Biol* 194, 165–175.

Chasiotis, H., Ionescu, A., Misyura, L., Bui, P., Fazio, K., Wang, J., Patrick, M., Weihrauch, D., Donini, A., 2016. An animal homolog of plant Mep/Amt transporters promotes ammonia excretion by the anal papillae of the disease vector mosquito *Aedes aegypti*. *J Exp Biol* 219, 1346–1355.

Cheng, H.T., Kim, M., Valerius, M.T., Surendran, K., Schuster-Gossler, K., Gossler, A., McMahon, A.P., Kopan, R., 2007. Notch2, but not Notch1, is required for proximal fate acquisition in the mammalian nephron. *Development* 134, 801–811.

Chitwood, B.G., Chitwood, M.B., 1950. An introduction to nematology. Section I. Anatomy. *Science* 113, 17.

Cima, F., Caicci, F., Sordino, P., 2014. The haemocytes of the salp *Thalia democratica* (Tunicata, Thaliacea): An ultrastructural and histochemical study in the oozoid. *Acta Zool* 95, 375–391.

Coons, L.B., L'Amoreaux, W.J., Rosell-Davis, R., Starr-spires, L., 1990. Fine structure of the fat body and nephrocytes in the life-stages of *Dermacentor variabilis*. *Exp Appl Acarol* 8, 125–142.

Costantini, F., Shakya, R., 2006. GDNF/Ret signaling and the development of the kidney. *Bioessays* 28, 117–127.

Crossley, A.C., 1972. The ultrastructure and function of pericardial cells and other nephrocytes in an insect: *Calliphora erythrocephala*. *Tissue Cell* 4, 529–560.

Denholm, B., Sudarsan, V., Pasalodos-Sanchez, S., Artero, R., Lawrence, P., Maddrell, S., Baylies, M., Skaer, H., 2003. Dual origin of the renal tubules in *Drosophila*: Mesodermal cells integrate and polarize to establish secretory function. *Curr Biol* 13, 1052–1057.

Desgrange, A., Cereghini, S., 2015. Nephron patterning: Lessons from *Xenopus*, zebrafish, and mouse studies. *Cells* 4, 483–499.

Dilly, P.N., Welsch, U., Rehkamper, G., 1986. Fine-structure of heart, pericardium and glomerular vessel in *Cephalodiscus gracilis* Mintosh, 1882 (Pterobranchia, Hemichordata). *Acta Zool* 67, 173–179.

Dobbs, G.H., Devries, A.L., 1975. Aglomerular nephron of antarctic teleosts – Light and electron-microscopic study. *Tissue Cell* 7, 159–170.

Dow, J.A., Davies, S.A., 2001. The *Drosophila melanogaster* malpighian tubule. *Adv Insect Physiol* 28, 1–83.

Dow, J.A., Maddrell, S.H., Gortz, A., Skaer, N.J., Brogan, S., Kaiser, K., 1994. The malpighian tubules of *Drosophila melanogaster*: A novel phenotype for studies of fluid secretion and its control. *J Exp Biol* 197, 421–428.

Dressler, G.R., 2006. The cellular basis of kidney development. *Annu Rev Cell Dev Biol* 22, 509–529.

Dube, K., McDonald, D.G., O'Donnell, M.J., 2000. Calcium transport by isolated anterior and posterior Malpighian tubules of *Drosophila melanogaster*: Roles of sequestration and secretion. *J Insect Physiol* 46, 1449–1460.

Durant, A.C., Donini, A., 2018. Evidence that Rh proteins in the anal papillae of the freshwater mosquito *Aedes aegypti* are involved in the regulation of acid-base balance in elevated salt and ammonia environments. *J Exp Biol* 221, jeb186866.

Ecelbarger, C.A., Terris, J., Frindt, G., Echevarria, M., Marples, D., Nielsen, S., Knepper, M.A., 1995. Aquaporin-3 water channel localization and regulation in rat kidney. *Am J Physiol* 269, F663–F672.

Ehlers, U., 1992. Dermonephridia – Modified epidermal cells with a probable excretory function in *Paratomella rubra* (Acoela, Plathelminthes). *Microfauna Marina* 7, 253–264.

Eichelberg, D., Wessing, A., 1975. Morphology of the Malpighian tubules of insects. *Fortschr Zool* 23, 124–147.

El-Shoura, S., 1989. Ultrastructure of the larval haemocytes and nephrocytes in the tick *Ornithodoros (Pavlovoskyella) erraticus* (Ixodoidea: Argasidae). *Acarologia* 30, 35–40.

Elkjaer, M.L., Nejsum, L.N., Gresz, V., Kwon, T.H., Jensen, U.B., Frokiaer, J., Nielsen, S., 2001. Immunolocalization of aquaporin-8 in rat kidney, gastrointestinal tract, testis, and airways. *Am J Physiol Renal Physiol* 281, F1047–F1057.

Engelund, M.B., Madsen, S.S., 2015. Tubular localization and expressional dynamics of aquaporins in the kidney of seawater-challenged Atlantic salmon. *J Comp Physiol B* 185, 207–223.

Evans, J.M., Allan, A.K., Davies, S.A., Dow, J.A., 2005. Sulphonylurea sensitivity and enriched expression implicate inward rectifier K^+ channels in *Drosophila melanogaster* renal function. *J Exp Biol* 208, 3771–3783.

Ezhova, O.V., Lavrova, E.A., Malakhov, V.V., 2013. Microscopic anatomy of the axial complex in the starfish *Asterias rubens* (Echinodermata, Asteroidea). *Biol. Bull.* 40, 643–653.

Ezhova, O.V., Lavrova, E.A., Malakhov, V.V., 2014. The morphology of the axial complex and associated structures in Asterozoa (Asteroidea, Echinoidea, Ophiuroidea). *Biologiya Morya (Vladivostok)* 40, 165–177.

Flessner, M.F., Wall, S.M., Knepper, M.A., 1991. Permeabilities of rat collecting duct segments to NH3 and NH4+. *Am J Physiol* 260, F264–F272.

Fushimi, K., Uchida, S., Hara, Y., Hirata, Y., Marumo, F., Sasaki, S., 1993. Cloning and expression of apical membrane water channel of rat kidney collecting tubule. *Nature* 361, 549–552.

Garayoa, M., Villaro, A.C., Montuenga, L., Sesma, P., 1992. Malpighian tubules of *Formica polyctena* (Hymenoptera): Light and electron microscopic study. *J Morphol* 214, 159–171.

Generlich, O., Giere, O., 1996. Osmoregulation in two aquatic oligochaetes from habitats with different salinity and comparison to other annelids. *Aquatic Oligochaete Biology VI*, Springer, Berlin, pp. 251–261.

George, W.C., 1939. Memoirs: A comparative study of the blood of the tunicates. *J Cell Sci* 2, 391–428.

Gérard, P., 1936. Comparative histophysiology of the vertebrate nephron. *J Anat* 70, 354.

Gerke, P., Huber, T.B., Sellin, L., Benzing, T., Walz, G., 2003. Homodimerization and heterodimerization of the glomerular podocyte proteins nephrin and NEPH1. *J Am Soc Nephrol* 14, 918–926.

Gomes, D., Agasse, A., Thiebaud, P., Delrot, S., Geros, H., Chaumont, F., 2009. Aquaporins are multifunctional water and solute transporters highly divergent in living organisms. *Biochim Biophys Acta* 1788, 1213–1228.

Gong, K.Q., Yallowitz, A.R., Sun, H., Dressler, G.R., Wellik, D.M., 2007. A Hox-Eya-Pax complex regulates early kidney developmental gene expression. *Mol Cell Biol* 27, 7661–7668.

Good, D.W., 1994. Ammonium transport by the thick ascending limb of Henle's loop. *Annu Rev Physiol* 56, 623–647.

Good, D.W., Knepper, M.A., Burg, M.B., 1984. Ammonia and bicarbonate transport by thick ascending limb of rat kidney. *Am J Physiol* 247, F35–F44.
Goodrich, E.S., 1945. The study of Nephridia and genital ducts since 1895. *J Cell Sci* 2, 113–301.
Green, L.F.B., 1979. Regional Specialization in the malpighian tubules of the New Zealand glow worm *Arachnocampa luminosa* (Diptera, Mycetophilidae) – Structure and function of type-I and type-II cells. *Tissue Cell* 11, 673–702.
Green, L.F.B., 1980. Cryptonephric malpighian tubule system in a dipteran larva, the New Zealand glow worm, *Arachnocampa luminosa* (Diptera, Mycetophilidae) – Structural study. *Tissue Cell* 12, 141–151.
Greven, H., 1982. Homologues or analogues? A survey of some structural patterns in Tardigrada, *Proceedings of the Third International Symposium on the Tardigrada* (Nelson, D.R. Ed.), pp. 55–76.
Haldin, C.E., Masse, K.L., Bhamra, S., Simrick, S., Kyuno, J., Jones, E.A., 2008. The lmx1b gene is pivotal in glomus development in *Xenopus laevis*. *Dev Biol* 322, 74–85.
Hasse, C., Rebscher, N., Reiher, W., Sobjinski, K., Moerschel, E., Beck, L., Tessmar-Raible, K., Arendt, D., Hassel, M., 2010. Three consecutive generations of nephridia occur during development of *Platynereis dumerilii* (Annelida, Polychaeta). *Dev Dyn* 239, 1967–1976.
Haszprunar, G., 1996. The molluscan rhogocyte (pore-cell, Blasenzelle, cellule nucale), and its significance for ideas on nephridial evolution. *J Molluscan Stud* 62, 185–211.
Hatton-Ellis, E., Ainsworth, C., Sushama, Y., Wan, S., VijayRaghavan, K., Skaer, H., 2007. Genetic regulation of patterned tubular branching in *Drosophila*. *Proc Natl Acad Sci USA* 104, 169–174.
Hay-Schmidt, A., 1987. The ultrastructure of the protonephridium of the actinotroch larva (Phoronida). *Acta Zool (Copenhagen)* 68, 35–47.
Hazelton, S.R., Townsend, V.R., Richter, C., Ritter, M.E., Felgenhauer, B.E., Spring, J.H., 2002. Morphology and ultrastructure of the malpighian tubules of the Chilean common tarantula (Araneae: Theraphosidae). *J Morphol* 251, 73–82.
Hediger, M.A., Clemencon, B., Burrier, R.E., Bruford, E.A., 2013. The ABCs of membrane transporters in health and disease (SLC series): introduction. *Mol Aspects Med* 34, 95–107.
Heliot, C., Desgrange, A., Buisson, I., Prunskaite-Hyyrylainen, R., Shan, J., Vainio, S., Umbhauer, M., Cereghini, S., 2013. HNF1B controls proximal-intermediate nephron segment identity in vertebrates by regulating Notch signalling components and Irx1/2. *Development* 140, 873–885.
Hessler, R.R., Elofsson, R., 1995. Segmental podocytic excretory glands in the thorax of *Hutchinsoniella macracantha* (cephalocarida). *J Crustacean Biol* 15, 61–69.
Holterman, M., van der Wurff, A., van den Elsen, S., van Megen, H., Bongers, T., Holovachov, O., Bakker, J., Helder, J., 2006. Phylum-wide analysis of SSU rDNA reveals deep phylogenetic relationships among nematodes and accelerated evolution toward crown clades. *Mol Biol Evol* 23, 1792–1800.
Ichimura, K., Sakai, T., 2017. Evolutionary morphology of podocytes and primary urine-producing apparatus. *Anat Sci Int* 92, 161–172.
Jack, J., Myette, G., 1999. Mutations that alter the morphology of the malpighian tubules in *Drosophila*. *Dev Genes Evol* 209, 546–554.
James, R.G., Kamei, C.N., Wang, Q., Jiang, R., Schultheiss, T.M., 2006. Odd-skipped related 1 is required for development of the metanephric kidney and regulates formation and differentiation of kidney precursor cells. *Development* 133, 2995–3004.
James, R.G., Schultheiss, T.M., 2005. Bmp signaling promotes intermediate mesoderm gene expression in a dose-dependent, cell-autonomous and translation-dependent manner. *Dev Biol* 288, 113–125.

Johnson, I.T., Riegel, J.A., 1977. Ultrastructural studies on malpighian tubule of pill millipede, *Glomeris marginata* (villers) – General morphology and localization of phosphatase enzymes. *Cell Tissue Res* 180, 357–366.

Jondelius, U., Ruiz-Trillo, I., Baguna, J., Riutort, M., 2002. The Nemertodermatida are basal bilaterians and not members of the Platyhelminthes. *Zool Scripta* 31, 201–215.

Kapoor, N.N., 1994. A study of the Malpighian tubules of the plecopteran nymph *Paragnetina media* (walker) (Plecoptera, Perlidae) by light, scanning electron, and transmission electron microscopy. *Canad J Zool-Rev Canad Zool* 72, 1566–1575.

Kaufmann, N., Mathai, J.C., Hill, W.G., Dow, J.A., Zeidel, M.L., Brodsky, J.L., 2005. Developmental expression and biophysical characterization of a *Drosophila melanogaster* aquaporin. *Am J Physiol Cell Physiol* 289, C397–C407.

Khodabandeh, S., Charmantier, G., Blasco, C., Grousset, E., Charmantier-Daures, M., 2005. Ontogeny of the antennal glands in the crayfish *Astacus leptodactylus* (Crustacea, Decapoda): anatomical and cell differentiation. *Cell Tissue Res* 319, 153–165.

Kieneke, A., Ahlrichs, W.H., Arbizu, P.M., Bartolomaeus, T., 2008. Ultrastructure of protonephridia in *Xenotrichula carolinensis syltensis* and *Chaetonotus maximus* (Gastrotricha: Chaetonotida): Comparative evaluation of the gastrotrich excretory organs. *Zoomorphology* 127, 1–20.

Kieneke, A., Hochberg, R., 2012. Ultrastructural observations of the protonephridia of *Polymerurus nodicaudus* (Gastrotricha: Paucitubulatina). *Acta Zool* 93, 115–124.

Kim, Y.K., Lee, S.Y., Kim, B.S., Kim, D.S., Nam, Y.K., 2014. Isolation and mRNA expression analysis of aquaporin isoforms in marine medaka *Oryzias dancena*, a euryhaline teleost. *Comp Biochem Physiol A Mol Integr Physiol* 171, 1–8.

King, B., Denholm, B., 2014. Malpighian tubule development in the red flour beetle (*Tribolium castaneum*). *Arthropod Struct Dev* 43, 605–613.

Kispert, A., Vainio, S., McMahon, A.P., 1998. Wnt-4 is a mesenchymal signal for epithelial transformation of metanephric mesenchyme in the developing kidney. *Development* 125, 4225–4234.

Knepper, M.A., Packer, R., Good, D.W., 1989. Ammonium transport in the kidney. *Physiol Rev* 69, 179–249.

Kobayashi, A., Valerius, M.T., Mugford, J.W., Carroll, T.J., Self, M., Oliver, G., McMahon, A.P., 2008. Six2 defines and regulates a multipotent self-renewing nephron progenitor population throughout mammalian kidney development. *Cell Stem Cell* 3, 169–181.

Koch, M., Quast, B., Bartolomaeus, T., 2014. *Coeloms and Nephridia in Annelids and Arthropods. Deep Metazoan Phylogeny: The Backbone of the Tree of Life. New Insights from Analyses of Molecules, Morphology, and Theory of Data Analysis*. Berlin: De Gruyter, pp. 173–284.

Kokkinopoulou, M., Guler, M.A., Lieb, B., Barbeck, M., Ghanaati, S., Markl, J., 2014. 3D-ultrastructure, functions and stress responses of gastropod (*Biomphalaria glabrata*) rhogocytes. *PLoS One* 9, e101078.

Kokkinopoulou, M., Spiecker, L., Messerschmidt, C., Barbeck, M., Ghanaati, S., Landfester, K., Markl, J., 2015. On the Ultrastructure and function of rhogocytes from the pond snail *Lymnaea stagnalis*. *PLoS One* 10, e0141195.

Kolosov, D., O'Donnell, M.J., 2020. Mechanisms and regulation of chloride transport in the Malpighian tubules of the larval cabbage looper *Trichoplusia ni*. *Insect Biochem Mol Biol* 116, 103263.

Koriyama, Y., Yamada, E., Watanabe, I., 1992. Pored-domes of the fenestrated endotheliocyte of the glomerular and peritubular capillaries in the rodent kidney. *J Electr Microsc* 41, 30–36.

Kozmik, Z., Holland, N.D., Kalousova, A., Paces, J., Schubert, M., Holland, L.Z., 1999. Characterization of an amphioxus paired box gene, AmphiPax2/5/8: Developmental expression patterns in optic support cells, nephridium, thyroid-like structures and pharyngeal gill slits, but not in the midbrain-hindbrain boundary region. *Development* 126, 1295–1304.

Kreidberg, J.A., Sariola, H., Loring, J.M., Maeda, M., Pelletier, J., Housman, D., Jaenisch, R., 1993. WT-1 is required for early kidney development. *Cell* 74, 679–691.

Krishnamoorthi, B., 1963. Gross morphology and histology of nephridia in four species of polychaetes, *Proceedings of the Indian Academy of Sciences Section B*, Springer, India, pp. 195–208.

Kristensen, R.M., Hay-Schmidt, A., 1989. The protonephridia of the arctic kinorhynch *Echinoderes aquilonius* (Cyclorhagida, Echinoderidae). *Acta Zool* 70, 13–27.

Kume, T., Deng, K., Hogan, B.L., 2000. Murine forkhead/winged helix genes Foxc1 (Mf1) and Foxc2 (Mfh1) are required for the early organogenesis of the kidney and urinary tract. *Development* 127, 1387–1395.

Kuzmina, T.V., Malakhov, V.V., 2015. The accessory hearts of the articulate brachiopod *Hemithyris psittacea*. *Zoomorphology* 134, 25–32.

Lammert, V., 1985. The fine-structure of protonephridia in gnathostomulida and their comparison within bilateria. *Zoomorphology* 105, 308–316.

Landowski, C.P., Suzuki, Y., Hediger, M.A., 2008. *The Mammalian Transporter Families, Seldin and Giebisch's the Kidney*. Amsterdam: Elsevier, pp. 91–146.

Larsen, E.H., Deaton, L.E., Onken, H., O'Donnell, M., Grosell, M., Dantzler, W.H., Weihrauch, D., 2014. Osmoregulation and excretion. *Compr Physiol* 4, 405–573.

Lavery, D.L., Davenport, I.R., Turnbull, Y.D., Wheeler, G.N., Hoppler, S., 2008. Wnt6 expression in epidermis and epithelial tissues during *Xenopus* organogenesis. *Dev Dyn* 237, 768–779.

Lee, S., Lee, H.J., Yang, H.S., Thornell, I.M., Bevensee, M.O., Choi, I., 2010. Sodium-bicarbonate cotransporter NBCn1 in the kidney medullary thick ascending limb cell line is upregulated under acidic conditions and enhances ammonium transport. *Exp Physiol* 95, 926–937.

Li, Q.L., Zhong, H.Y., Zhang, Y.L., Wei, C., 2015. Comparative morphology of the distal segments of Malpighian tubules in cicadas and spittlebugs, with reference to their functions and evolutionary indications to Cicadomorpha (Hemiptera: Auchenorrhyncha). *Zool Anz* 258, 54–68.

Li, Y., Cheng, C.N., Verdun, V.A., Wingert, R.A., 2014. Zebrafish nephrogenesis is regulated by interactions between retinoic acid, mecom, and Notch signaling. *Dev Biol* 386, 111–122.

Lindström, N.O., Tran, T., Guo, J.J., Rutledge, E., Parvez, R.K., Thornton, M.E., Grubbs, B., McMahon, J.A., McMahon, A.P., 2018. Conserved and divergent molecular and anatomic features of human and mouse nephron patterning. *J Am Soc Nephrol* 29, 825–840.

Linser, P.J., Neira Oviedo, M., Hirata, T., Seron, T.J., Smith, K.E., Piermarini, P.M., Romero, M.F., 2012. Slc4-like anion transporters of the larval mosquito alimentary canal. *J Insect Physiol* 58, 551–562.

Little, M., Georgas, K., Pennisi, D., Wilkinson, L., 2010. *Kidney Development: Two Tales of Tubulogenesis, Current Topics in Developmental Biology*. Amsterdam: Elsevier, pp. 193–229.

Lüter, C., 1995. Ultrastructure of the metanephridia of *Terebratulina retusa* and *Crania anomala* (Brachiopoda). *Zoomorphology* 115, 99–107.

Ma, M., Jiang, Y.J., 2007. Jagged2a-notch signaling mediates cell fate choice in the zebrafish pronephric duct. *PLoS Genet* 3, e18.

Maddrell, S., 1980. Characteristics of epithelial transport in insect Malpighian tubules. In: *Current Topics in Membranes and Transport*. Amsterdam: Elsevier, pp. 427–463.

Madsen, S.S., Bujak, J., Tipsmark, C.K., 2014. Aquaporin expression in the Japanese medaka (*Oryzias latipes*) in freshwater and seawater: challenging the paradigm of intestinal water transport? *J Exp Biol* 217, 3108–3121.

Madsen, S.S., Olesen, J.H., Bedal, K., Engelund, M.B., Velasco-Santamaria, Y.M., Tipsmark, C.K., 2011. Functional characterization of water transport and cellular localization of three aquaporin paralogs in the salmonid intestine. *Front Physiol* 2, 56.

Majumdar, A., Lun, K., Brand, M., Drummond, I.A., 2000. Zebrafish no isthmus reveals a role for pax2.1 in tubule differentiation and patterning events in the pronephric primordia. *Development* 127, 2089–2098.

Mak, D.O., Dang, B., Weiner, I.D., Foskett, J.K., Westhoff, C.M., 2006. Characterization of ammonia transport by the kidney Rh glycoproteins RhBG and RhCG. *Am J Physiol Renal Physiol* 290, F297–F305.

Marx, V., 2019. A dream of single-cell proteomics. *Nat Methods* 16, 809–812.

Mayer, G., 2006. Origin and differentiation of nephridia in the Onychophora provide no support for the Articulata. *Zoomorphology* 125, 1–12.

Mayer, G., Bartolomaeus, T., 2003. Ultrastructure of the stomochord and the heart-glomerulus complex in *Rhabdopleura compacta* (Pterobranchia): Phylogenetic implications. *Zoomorphology* 122, 125–133.

McLaren, D.J., 1974. The anterior glands of adult *Necator americanus* (Nematoda: Strongyloidea). I. Ultrastructural studies. *Int J Parasitol* 4, 25–37.

McLaughlin, K.A., Rones, M.S., Mercola, M., 2000. Notch regulates cell fate in the developing pronephros. *Dev Biol* 227, 567–580.

McMahon, A.P., 2016. *Development of the Mammalian Kidney, Current topics in Developmental Biology*. Amsterdam: Elsevier, pp. 31–64.

Miner, J.H., Morello, R., Andrews, K.L., Li, C., Antignac, C., Shaw, A.S., Lee, B., 2002. Transcriptional induction of slit diaphragm genes by Lmx1b is required in podocyte differentiation. *J Clin Invest* 109, 1065–1072.

Miyaki, T., Kawasaki, Y., Matsumoto, A., Kakuta, S., Sakai, T., & Ichimura, K. (2020). Nephrocytes are part of the spectrum of filtration epithelial diversity. *Cell and Tissue Research*, 382(3), 609–625.

Møbjerg, N., Dahl, C., 1996. Studies on the morphology and ultrastructure of the Malpighian tubules of *Halobiotus crispae* Kristensen, 1982 (Eutardigrada). *Zool J Linnean Soc* 116, 85–99.

Møbjerg, N., Larsen, E.H., Jespersen, A., 2000. Morphology of the kidney in larvae of *Bufo viridis* (Amphibia, Anura, Bufonidae). *J Morphol* 245, 177–195.

Moller, P.C., Ellis, R.A., 1974. Fine structure of excretory system of amphioxus (*Branchiostoma floridae*) and its response to osmotic stress. *Cell Tissue Res* 148, 1–9.

Moore, M.W., Klein, R.D., Farinas, I., Sauer, H., Armanini, M., Phillips, H., Reichardt, L.F., Ryan, A.M., Carver-Moore, K., Rosenthal, A., 1996. Renal and neuronal abnormalities in mice lacking GDNF. *Nature* 382, 76–79.

Moriguchi, T., Hamada, M., Morito, N., Terunuma, T., Hasegawa, K., Zhang, C., Yokomizo, T., Esaki, R., Kuroda, E., Yoh, K., Kudo, T., Nagata, M., Greaves, D.R., Engel, J.D., Yamamoto, M., Takahashi, S., 2006. MafB is essential for renal development and F4/80 expression in macrophages. *Mol Cell Biol* 26, 5715–5727.

Morishita, Y., Matsuzaki, T., Hara-chikuma, M., Andoo, A., Shimono, M., Matsuki, A., Kobayashi, K., Ikeda, M., Yamamoto, T., Verkman, A., Kusano, E., Ookawara, S., Takata, K., Sasaki, S., Ishibashi, K., 2005. Disruption of aquaporin-11 produces polycystic kidneys following vacuolization of the proximal tubule. *Mol Cell Biol* 25, 7770–7779.

Mudumana, S.P., Hentschel, D., Liu, Y., Vasilyev, A., Drummond, I.A., 2008. Odd skipped related1 reveals a novel role for endoderm in regulating kidney versus vascular cell fate. *Development* 135, 3355–3367.

Mugford, J.W., Yu, J., Kobayashi, A., McMahon, A.P., 2009. High-resolution gene expression analysis of the developing mouse kidney defines novel cellular compartments within the nephron progenitor population. *Dev Biol* 333, 312–323.

Nakai, S., Sugitani, Y., Sato, H., Ito, S., Miura, Y., Ogawa, M., Nishi, M., Jishage, K., Minowa, O., Noda, T., 2003. Crucial roles of Brn1 in distal tubule formation and function in mouse kidney. *Development* 130, 4751–4759.

Naylor, R.W., Przepiorski, A., Ren, Q., Yu, J., Davidson, A.J., 2013. HNF1beta is essential for nephron segmentation during nephrogenesis. *J Am Soc Nephrol* 24, 77–87.

Needham, J., 1935. Problems of nitrogen catabolism in invertebrates: Correlation between uricotelic metabolism and habitat in the phylum Mollusca. *Biochem J* 29, 238.

Nejsum, L.N., Elkjaer, M., Hager, H., Frokiaer, J., Kwon, T.H., Nielsen, S., 2000. Localization of aquaporin-7 in rat and mouse kidney using RT-PCR, immunoblotting, and immunocytochemistry. *Biochem Biophys Res Commun* 277, 164–170.

Nelson, F.K., Albert, P.S., Riddle, D.L., 1983. Fine structure of the *Caenorhabditis elegans* secretory excretory system. *J Ultrastruct Res* 82, 156–171.

Neuhaus, B., 1988. Ultrastructure of the protonephridia in *Pycnophyes kielensis* (Kinorhyncha, Homalorhagida). *Zoomorphology* 108, 245–253.

Neuhaus, B., Kristensen, R.M., 2007. Ultrastructure of the protonephridia of larval *Rugiloricus* cf. *cauliculus*, male *Armorioricus elegans*, and female *Nanaloricus mysticus* (Loricifera). *J Morphol* 268, 357–370.

Nicholls, S.P., 1983. Ultrastructural evidence for paracellular fluid-flow in the malpighian tubules of a larval mayfly. *Tissue Cell* 15, 627–637.

Nielsen, S., Smith, B.L., Christensen, E.I., Knepper, M.A., Agre, P., 1993. CHIP28 water channels are localized in constitutively water-permeable segments of the nephron. *J Cell Biol* 120, 371–383.

Nishinakamura, R., Matsumoto, Y., Nakao, K., Nakamura, K., Sato, A., Copeland, N.G., Gilbert, D.J., Jenkins, N.A., Scully, S., Lacey, D.L., Katsuki, M., Asashima, M., Yokota, T., 2001. Murine homolog of SALL1 is essential for ureteric bud invasion in kidney development. *Development* 128, 3105–3115.

Nishinakamura, R., Sakaguchi, M., 2014. BMP signaling and its modifiers in kidney development. *Pediatr Nephrol* 29, 681–686.

Nocelli, R., Cintra-Socolowski, P., Roat, T., Silva-Zacarin, E., Malaspina, O., 2016. Comparative physiology of Malpighian tubules: Form and function. *Insect Physiol* 2016, 13–23.

O'Brien, L.L., Grimaldi, M., Kostun, Z., Wingert, R.A., Selleck, R., Davidson, A.J., 2011. Wt1a, Foxc1a, and the Notch mediator Rbpj physically interact and regulate the formation of podocytes in zebrafish. *Dev Biol* 358, 318–330.

O'Connor, K.R., Beyenbach, K.W., 2001. Chloride channels in apical membrane patches of stellate cells of Malpighian tubules of *Aedes aegypti*. *J Exp Biol* 204, 367–378.

O'Donnell, M., 2010. Mechanisms of excretion and ion transport in invertebrates. *Compr Physiol* 1207–1289.

O'Donnell, M.J., Maddrell, S.H., 1995. Fluid reabsorption and ion transport by the lower Malpighian tubules of adult female *Drosophila*. *J Exp Biol* 198, 1647–1653.

O'Donnell, M.J., Rheault, M.R., Davies, S.A., Rosay, P., Harvey, B.J., Maddrell, S.H., Kaiser, K., Dow, J.A., 1998. Hormonally controlled chloride movement across *Drosophila* tubules is via ion channels in stellate cells. *Am J Physiol* 274, R1039–R1049.

Pal, R., Kumar, K., 2012. Ultrastructural features of the larval Malpighian tubules of the flesh fly *Sarcophaga ruficornis* (Diptera: Sarcophagidae). *Int J Trop Insect Sci* 32, 166–172.

Pandey, R.N., Yaganti, S., Coffey, S., Frisbie, J., Alnajjar, K., Goldstein, D., 2010. Expression and immunolocalization of aquaporins HC-1, -2, and -3 in Cope's gray treefrog, *Hyla chrysoscelis*. *Comp Biochem Physiol A Mol Integr Physiol* 157, 86–94.

Patari-Sampo, A., Ihalmo, P., Holthofer, H., 2006. Molecular basis of the glomerular filtration: Nephrin and the emerging protein complex at the podocyte slit diaphragm. *Ann Med* 38, 483–492.

Paulus, H.F., 2000. Phylogeny of the myriapoda – crustacea – insecta: A new attempt using photoreceptor structure. *J Zool Syst Evol Res* 38, 189–208.

Perner, B., Englert, C., Bollig, F., 2007. The Wilms tumor genes wt1a and wt1b control different steps during formation of the zebrafish pronephros. *Dev Biol* 309, 87–96.

Peters, W., 1977. Possible sites of ultrafiltration in *Tubifex tubifex* muller (Annelida, Oligochaeta). *Cell Tissue Res* 179, 367–375.

Pfeffer, P.L., Gerster, T., Lun, K., Brand, M., Busslinger, M., 1998. Characterization of three novel members of the zebrafish Pax2/5/8 family: Dependency of Pax5 and Pax8 expression on the Pax2.1 (noi) function. *Development* 125, 3063–3074.

Piermarini, P.M., Grogan, L.F., Lau, K., Wang, L., Beyenbach, K.W., 2010. A SLC4-like anion exchanger from renal tubules of the mosquito (*Aedes aegypti*): Evidence for a novel role of stellate cells in diuretic fluid secretion. *Am J Physiol Regul Integr Comp Physiol* 298, R642–R660.

Preisig, P.A., Alpern, R.J., 1990. Pathways for apical and basolateral membrane NH_3 and NH_4^+ movement in rat proximal tubule. *Am J Physiol* 259, F587–F593.

Pullikuth, A.K., Aimanova, K., Kang'ethe, W., Sanders, H.R., Gill, S.S., 2006. Molecular characterization of sodium/proton exchanger 3 (NHE3) from the yellow fever vector, *Aedes aegypti*. *J Exp Biol* 209, 3529–3544.

Quaggin, S.E., Kreidberg, J.A., 2008. Development of the renal glomerulus: Good neighbors and good fences. *Development* 135, 609–620.

Quijada-Rodriguez, A.R., Treberg, J.R., Weihrauch, D., 2015. Mechanism of ammonia excretion in the freshwater leech *Nephelopsis obscura*: Characterization of a primitive Rh protein and effects of high environmental ammonia. *Am J Physiol Regul Integr Comp Physiol* 309, R692–R705.

Raciti, D., Reggiani, L., Geffers, L., Jiang, Q., Bacchion, F., Subrizi, A.E., Clements, D., Tindal, C., Davidson, D.R., Kaissling, B., Brandli, A.W., 2008. Organization of the pronephric kidney revealed by large-scale gene expression mapping. *Genome Biol* 9, R84.

Randsø, P.V., Yamasaki, H., Bownes, S.J., Herranz, M., Di Domenico, M., Qii, G.B., Sørensen, M.V., 2019. Phylogeny of the *Echinoderes coulli*-group (Kinorhyncha: Cyclorhagida: Echinoderidae) – A cosmopolitan species group trapped in the intertidal. *Invertebr Syst* 33, 501–517.

Rheault, M.R., Okech, B.A., Keen, S.B., Miller, M.M., Meleshkevitch, E.A., Linser, P.J., Boudko, D.Y., Harvey, W.R., 2007. Molecular cloning, phylogeny and localization of AgNHA1: The first Na^+/H^+ antiporter (NHA) from a metazoan, *Anopheles gambiae*. *J Exp Biol* 210, 3848–3861.

Riemann, O., Ahlrichs, W.H., 2010. The evolution of the protonephridial terminal organ across Rotifera with particular emphasis on *Dicranophorus forcipatus*, *Encentrum mucronatum* and *Erignatha clastopis* (Rotifera: Dicranophoridae). *Acta Zool (Copenhagen)* 91, 199–211.

Rivest, B.R., 1992. Studies on the structure and function of the larval kidney complex of prosobranch gastropods. *Biol Bull* 182, 305–323.

Rohde, K., 1991. The evolution of protonephridia of the Platyhelminthes. *Hydrobiologia* 227, 315–321.

Rohde, K., Cannon, L.R., Watson, N., 1988. Ultrastructure of the protonephridia of *Monocelis* (Proseriata, Monocelididae). *J Submicrosc Cytol Pathol* 20, 425–435.

Rohde, K., Watson, N.A., 1993. Ultrastructure of the protonephridial system of regenerating *Stenostomum* sp. (Platyhelminthes, Catenulida). *Zoomorphology (Berlin)* 113, 61–67.

Ruppert, E.E., 1994. Evolutionary origin of the vertebrate nephron. *Am Zool* 34, 542–553.

Ruppert, E.E., Smith, P.R., 1988. The functional organization of filtration nephridia. *Biol Rev Camb Philos Soc* 63, 231–258.
Ruthensteiner, B., Wanninger, A., Haszprunar, G., 2001. The protonephridial system of the tusk shell, *Antalis entalis* (Mollusca, Scaphopoda). *Zoomorphology (Berlin)* 121, 19–26.
Sajithlal, G., Zou, D., Silvius, D., Xu, P.X., 2005. Eya 1 acts as a critical regulator for specifying the metanephric mesenchyme. *Dev Biol* 284, 323–336.
Santos, C.R., Estevao, M.D., Fuentes, J., Cardoso, J.C., Fabra, M., Passos, A.L., Detmers, F.J., Deen, P.M., Cerda, J., Power, D.M., 2004. Isolation of a novel aquaglyceroporin from a marine teleost (*Sparus auratus*): Function and tissue distribution. *J Exp Biol* 207, 1217–1227.
Saulnier, D.M., Ghanbari, H., Brandli, A.W., 2002. Essential function of Wnt-4 for tubulogenesis in the *Xenopus* pronephric kidney. *Dev Biol* 248, 13–28.
Saxén, L., Saxén, L., 1987. *Organogenesis of the Kidney*. Cambridge: Cambridge University Press.
Schill, R.O., 2019. *Water Bears: The Biology of Tardigrades*. Berlin: Springer.
Schmidt-Nielsen, K., 1997. *Animal Physiology: Adaptation and Environment*. Cambridge: Cambridge University Press.
Schmidt-Rhaesa, A., 2007. *The Evolution of Organ Systems*. Oxford: Oxford University Press.
Schwaha, T.F., Ostrovsky, A.N., Wanninger, A., 2020. Key novelties in the evolution of the aquatic colonial phylum Bryozoa: Evidence from soft body morphology. *Biol Rev* 95, 696–729.
Scimone, M.L., Srivastava, M., Bell, G.W., Reddien, P.W., 2011. A regulatory program for excretory system regeneration in planarians. *Development* 138, 4387–4398.
Sebe-Pedros, A., Burkhardt, P., Sanchez-Pons, N., Fairclough, S.R., Lang, B.F., King, N., Ruiz-Trillo, I., 2013. Insights into the origin of metazoan filopodia and microvilli. *Mol Biol Evol* 30, 2013–2023.
Seifert, G., Rosenberg, J., 1977. Die Ultrastruktur der nephrozyten von *Peripatoides leuckarti* (Saenger 1869) (Onychophora, Peripatopsidae). *Zoomorphologie* 86, 169–181.
Seitz, K.-A., 1987. *Excretory Organs, Ecophysiology of Spiders*. Berlin: Springer, pp. 239–248.
Shafer, M.E.R., 2019. Cross-species analysis of single-cell transcriptomic data. *Front Cell Dev Biol* 7, 175.
Shatrov, A.B., 1998. The ultrastructure and possible functions of nephrocytes in the trombiculid mite *Hirsutiella zachvatkini* (Acariformes: Trombiculidae). *Exp Appl Acarol* 22, 1–16.
Shim, K., Blake, K.J., Jack, J., Krasnow, M.A., 2001. The *Drosophila* ribbon gene encodes a nuclear BTB domain protein that promotes epithelial migration and morphogenesis. *Development* 128, 4923–4933.
Si, L., Pan, L., Wang, H., Zhang, X., 2018. Identification of the role of Rh protein in ammonia excretion of the swimming crab *Portunus trituberculatus*. *J Exp Biol* 221, jeb184655.
Smith, R., 1970. Hypo-osmotic urine in *Nereis diversicolor*. *J Exp Biol* 53, 101–108.
Smith, R., 1984. The larval nephridia of the brackish-water polychaete, *Nereis diversicolor*. *J Morphol* 179, 273–289.
Smythe, A.B., Holovachov, O., Kocot, K.M., 2019. Improved phylogenomic sampling of free-living nematodes enhances resolution of higher-level nematode phylogeny. *BMC Evol Biol* 19, 121.
Sozen, M.A., Armstrong, J.D., Yang, M., Kaiser, K., Dow, J.A., 1997. Functional domains are specified to single-cell resolution in a *Drosophila* epithelium. *Proc Natl Acad Sci USA* 94, 5207–5212.

Star, R.A., Kurtz, I., Mejia, R., Burg, M.B., Knepper, M.A., 1987. Disequilibrium pH and ammonia transport in isolated perfused cortical collecting ducts. *Am J Physiol* 253, F1232–F1242.

Storch, V., Herrmann, K., 1978. Podocytes in blood-vessel linings of *Phoronis muelleri* (Phoronida, Tentaculata). *Cell Tissue Res* 190, 553–556.

Taylor, H.H., 1971a. Fine structure of type-2 cells in malpighian tubules of stick insect, *Carausius morosus*. *Z Zellforsch Mikrosk Anat* 122, 411.

Taylor, H.H., 1971b. Water and solute transport by malpighian tubules of stick insect, *Carausius morosus* – Normal ultrastructure of type-1 cells. *Z Zellforsch Mikrosk Anat* 118, 333.

Temereva, E.N., Malakhov, V.V., 2006. Development of excretory organs in *Phoronopsis harmeri* (Phoronida): From protonephridium to nephromixium. *Zool Zh* 85, 915–924.

Tena, J.J., Neto, A., de la Calle-Mustienes, E., Bras-Pereira, C., Casares, F., Gomez-Skarmeta, J.L., 2007. Odd-skipped genes encode repressors that control kidney development. *Dev Biol* 301, 518–531.

Terris, J., Ecelbarger, C.A., Marples, D., Knepper, M.A., Nielsen, S., 1995. Distribution of aquaporin-4 water channel expression within rat kidney. *Am J Physiol* 269, F775–F785.

Tetelin, S., Jones, E.A., 2010. Xenopus Wnt11b is identified as a potential pronephric inducer. *Dev Dyn* 239, 148–159.

Teuchert, G., 1973. Die feinstruktur des protonephridialsystems von *Turbanella cornuta* remane, einem marinen gastrotrich der ordnung macrodasyoidea. *Z Zellforsch Mikrosk Anat* 136, 277–289.

Thiel, D., Hugenschutt, M., Meyer, H., Paululat, A., Quijada-Rodriguez, A.R., Purschke, G., Weihrauch, D., 2017. Ammonia excretion in the marine polychaete *Eurythoe complanata* (Annelida). *J Exp Biol* 220, 425–436.

Thomsen, J., Himmerkus, N., Holland, N., Sartoris, F.J., Bleich, M., Tresguerres, M., 2016. Ammonia excretion in mytilid mussels is facilitated by ciliary beating. *J Exp Biol* 219, 2300–2310.

Todt, C., Wanninger, A., 2010. Of tests, trochs, shells, and spicules: development of the basal mollusk *Wirenia argentea* (Solenogastres) and its bearing on the evolution of trochozoan larval key features. *Front Zool* 7, 1–17.

Tomar, R., Mudumana, S.P., Pathak, N., Hukriede, N.A., Drummond, I.A., 2014. osr1 is required for podocyte development downstream of wt1a. *J Am Soc Nephrol* 25, 2539–2545.

Torrie, L.S., Radford, J.C., Southall, T.D., Kean, L., Dinsmore, A.J., Davies, S.A., Dow, J.A., 2004. Resolution of the insect ouabain paradox. *Proc Natl Acad Sci USA* 101, 13689–13693.

Trapnell, C., 2015. Defining cell types and states with single-cell genomics. *Genome Res* 25, 1491–1498.

Tryggvason, K., Wartiovaara, J., 2001. Molecular basis of glomerular permselectivity. *Curr Opin Nephrol Hypertens* 10, 543–549.

Tsang, T.E., Shawlot, W., Kinder, S.J., Kobayashi, A., Kwan, K.M., Schughart, K., Kania, A., Jessell, T.M., Behringer, R.R., Tam, P.P., 2000. Lim1 activity is required for intermediate mesoderm differentiation in the mouse embryo. *Dev Biol* 223, 77–90.

Tschopp, P., Tabin, C.J., 2017. Deep homology in the age of next-generation sequencing. *Philos Trans R Soc Lond B Biol Sci* 372, 20150475.

Urban, A.E., Zhou, X., Ungos, J.M., Raible, D.W., Altmann, C.R., Vize, P.D., 2006. FGF is essential for both condensation and mesenchymal-epithelial transition stages of pronephric kidney tubule development. *Dev Biol* 297, 103–117.

Van Campenhout, C., Nichane, M., Antoniou, A., Pendeville, H., Bronchain, O.J., Marine, J.C., Mazabraud, A., Voz, M.L., Bellefroid, E.J., 2006. Evi1 is specifically expressed in the distal tubule and duct of the *Xenopus* pronephros and plays a role in its formation. *Dev Biol* 294, 203–219.

Vandenbulcke, F., Grelle, C., Fabre, M.C., Descamps, M., 1998. Ultrastructural and autometallographic studies of the nephrocytes of *Lithobius forficatus* L.(Myriapoda, Chilopoda): Role in detoxification of cadmium and lead. *Int J Insect Morphol Embryol* 27, 111–120.

Vize, P.D., Seufert, D.W., Carroll, T.J., Wallingford, J.B., 1997. Model systems for the study of kidney development: Use of the pronephros in the analysis of organ induction and patterning. *Dev Biol* 188, 189–204.

von Nordheim, H., Schrader, A., 1994. Ultrastructure and functional morphology of the protonephridia and segmental fenestrated lacunae in *Protodrilus rubropharyngeus* (Polychaeta, Protodrilidae). *Helgoländer Meeresunters* 48, 467–485.

Vu, H.T.K., Rink, J.C., McKinney, S.A., McClain, M., Lakshmanaperumal, N., Alexander, R., Alvarado, A.S., 2015. Stem cells and fluid flow drive cyst formation in an invertebrate excretory organ. *Elife* 4, e07405.

Waddell, A., 1968. The excretory system of the kidney worm *Stephanurus dentatus* (Nematoda). *Parasitology* 58, 907–919.

Wall, B.J., Oschman, J.L., Schmidt, B.A., 1975. Morphology and function of Malpighian tubules and associated structures in the cockroach, *Periplaneta americana*. *J Morphol* 146, 265–306.

Wall, S.M., 1996. Ammonium transport and the role of the Na,K-ATPase. *Miner Electrolyte Metab* 22, 311–317.

Warner, F.D., 1969. Fine structure of protonephridia in rotifer *Asplanchna*. *J Ultrastruct Res* 29, 499–+.

Weavers, H., Prieto-Sanchez, S., Grawe, F., Garcia-Lopez, A., Artero, R., Wilsch-Brauninger, M., Ruiz-Gomez, M., Skaer, H., Denholm, B., 2009. The insect nephrocyte is a podocyte-like cell with a filtration slit diaphragm. *Nature* 457, 322–326.

Węglarska, B., 1980. Light and electron-microscopic studies on the excretory system of *Macrobiotus richtersi* Murray, 1911 (Eutardigrada). *Cell Tissue Res* 207, 171–182.

Weihrauch, D., 2006. Active ammonia absorption in the midgut of the Tobacco hornworm *Manduca sexta* L.: Transport studies and mRNA expression analysis of a Rhesus-like ammonia transporter. *Insect Biochem Mol Biol* 36, 808–821.

Weihrauch, D., Chan, A.C., Meyer, H., Doring, C., Sourial, M., O'Donnell, M.J., 2012. Ammonia excretion in the freshwater planarian *Schmidtea mediterranea*. *J Exp Biol* 215, 3242–3253.

Weihrauch, D., Fehsenfeld, S., Quijada-Rodriguez, A., 2017. *Nitrogen Excretion in Aquatic Crustaceans, Acid–Base Balance and Nitrogen Excretion in Invertebrates*. Berlin: Springer, pp. 1–24.

Weihrauch, D., O'Donnell, M., 2017. *Acid-Base Balance and Nitrogen Excretion in Invertebrates*. Cham: Springer International Publishing, pp. 10, 978–973.

Wellik, D.M., Hawkes, P.J., Capecchi, M.R., 2002. Hox11 paralogous genes are essential for metanephric kidney induction. *Genes Dev* 16, 1423–1432.

Welsch, U., Rehkamper, G., 1987. Podocytes in the axial organ of echinoderms. *J Zool* 213, 45–50.

Wenning, A., Cahill, M.A., Greisinger, U., Kaltenhauser, U., 1993. Organogenesis in the leech: Development of nephridia, bladders and their innervation. *Roux Arch Dev Biol* 202, 329–340.

Werntz, H.O., 1963. Osmotic regulation in marine and fresh-water gammarids (Amphipoda). *Biol Bull* 124, 225–239.

Westheide, W., 1985. Ultrastructure of the protonephridia in the dorvilleid polychaete *Apodotrocha progenerans* (Annelida). *Zool Scripta* 14, 273–278.

White, J.T., Zhang, B., Cerqueira, D.M., Tran, U., Wessely, O., 2010. Notch signaling, wt1 and foxc2 are key regulators of the podocyte gene regulatory network in *Xenopus*. *Development* 137, 1863–1873.

Wigglesworth, V.B., 1943. The fate of haemoglobin in *Rhodnius prolixus* (Hemiptera) and other blood-sucking arthropods. *Proc R Soc Lond Ser B Biol Sci* 131, 313–339.

Wild, W., Pogge von Strandmann, E., Nastos, A., Senkel, S., Lingott-Frieg, A., Bulman, M., Bingham, C., Ellard, S., Hattersley, A.T., Ryffel, G.U., 2000. The mutated human gene encoding hepatocyte nuclear factor 1beta inhibits kidney formation in developing *Xenopus* embryos. *Proc Natl Acad Sci USA* 97, 4695–4700.

Wingert, R.A., Davidson, A.J., 2008. The zebrafish pronephros: A model to study nephron segmentation. *Kidney Int* 73, 1120–1127.

Wingert, R.A., Selleck, R., Yu, J., Song, H.D., Chen, Z., Song, A., Zhou, Y., Thisse, B., Thisse, C., McMahon, A.P., Davidson, A.J., 2007. The cdx genes and retinoic acid control the positioning and segmentation of the zebrafish pronephros. *PLoS Genet* 3, 1922–1938.

Wolff, J., Merker, H.-J., 1966. Ultrastruktur und bildung von poren im endothel von porösen und geschlossenen kapillaren. *Z Zellforsch Mikrosk Anat* 73, 174–191.

Woodruff, J.B., Mitchell, B.J., Shankland, M., 2007. Hau-Pax3/7A is an early marker of leech mesoderm involved in segmental morphogenesis, nephridial development, and body cavity formation. *Dev Biol* 306, 824–837.

Xu, P.X., Zheng, W., Huang, L., Maire, P., Laclef, C., Silvius, D., 2003. Six1 is required for the early organogenesis of mammalian kidney. *Development* 130, 3085–3094.

Yasui, M., Kwon, T.H., Knepper, M.A., Nielsen, S., Agre, P., 1999. Aquaporin-6: An intracellular vesicle water channel protein in renal epithelia. *Proc Natl Acad Sci USA* 96, 5808–5813.

Zhang, F., Zhao, Y., Chao, Y., Muir, K., Han, Z., 2013. Cubilin and amnionless mediate protein reabsorption in *Drosophila* nephrocytes. *J Am Soc Nephrol* 24, 209–216.

Zhuang, S., Shao, H., Guo, F., Trimble, R., Pearce, E., Abmayr, S.M., 2009. Sns and kirre, the drosophila orthologs of nephrin and Neph1, direct adhesion, fusion and formation of a slit diaphragm-like structure in insect nephrocytes. *Development* 136, 2335–2344.

Ziegler, A., Faber, C., Bartolomaeus, T., 2009. Comparative morphology of the axial complex and interdependence of internal organ systems in sea urchins (Echinodermata: Echinoidea). *Front Zool* 6, 10.

Index

A

Acoela, 61–62
Algoriphagus machipongonensis, 19
Ammonia, 147
Amniotes, 139
Amphimedon queenslandica, 3
Ancestral primordial germ cell, 56
Animal epithelia
    cell adhesion and polarity, 85–86
    genes encoding, 83–85
    integrins, 83–84
Animal multicellularity
    cell differentiation
        choanoblastea theory, 13–14
        spatial cell differentiation, 14
        sponges, 14
    transdifferentiation capacities
        Caenorhabditis elegans, 16
        choanocytes, 14
        choanoflagellates and metazoans, 14
        Dictyostelium discoideum, 16
        sponge choanocytes, 15
        tissue regeneration, 15
        unicellular holozoans, 16–21

B

Branchiostoma floridae, 144

C

Caenorhabditis elegans, 3, 16
Capsaspora owczarzaki, 18, 19, 21
Cell adhesion, 85–86
Cell theory, 101
Cell types
    cell diversity, 2
    choanocytes, 1
    classification, 1
    defined, 2
    evolutionary change, 6
    excretory cells, 139–140
    functional roles, 1
    gene regulatory network (GRN), 2
    global transcription patterns, 2
    homology problems, 4–5
    metanephridial systems, 136–139
    molecular characterization, 2
    morphology and function, 1
    protonephridia, 133–136
    secretory excretory organs, 132–133
    in sponges, 1
    stability of, 7
    steady renewal *vs.* terminal differentiation
        in *Amphimedon queenslandica*, 3
        in animals, 3
        in *Caenorhabditis elegans*, 3
        core gene regulatory network model, 3
        kernel model, 3
        molecular models, 3
        spermatogenesis, 3
        sponge archaeocytes, behavior of, 4
        terminal selector gene model, 3
    transcription factors, 2
Chelicerata, 132
Choanocytes, 14
Choanoeca flexa, 18
Cnidaria, 108
    gene expression, 59–60
    nematogenic machinery, function, and development, 40
    origin of gametes in, 58–59
Conventional embryogenesis, 48
Conventional high-throughput sequencing methods, 22
Creolimax fragrantissima, 18, 20
Crystal cells, 14
Ctenophora, 60–61
Ctenophores, 108

D

Dendrites, 105
Dictyostelium discoideum, 16
Dinoflagellates, 38, 40

E

Ectoderm, 130
Embryogenesis, 48
Endosymbiotic organelles, 28
Epithelia
    cell biology, 76
    cell polarity complexes and cell domains, 79–80
    classification of, 78
    control molecule and ion exchanges, 76
    definition of, 76–78
    epithelial regulatory signatures, 92
    of junction types, 78–79
    metazoan epithelium

adhesive junctions, 1
  ancestral basement membrane, 86–88
  communicating junctions, 89
  non-bilaterian relationships, 88
  sealing/occluding junctions, 89
metazoans, 80–82
multicellularity and epithelial-like structures, 76
Epithelial cells, 14
Euglena, 32
Eukaryotic genomics, 28
Eutardigrada, 132
Excretory cells, 139–140
Excretory organs, 132–133
  and adaptations, 140
  functional compartmentalization, 144–147
  insect Malpighian tubules, 144
  nephron development, 141–143
  nitrogenous waste excretion, 147–148
  protonephridial development, 143–144
Extracellular matrix (ECM), 131

F

Fiber cells, 14

G

GABAergic synapse, 106
Gastropod molluscs, 105
Gene expression, 6, 27, 59–60
Gene regulatory network (GRN), 2
Genes encoding, 83
Genetic transformation, 29
Geodia atlantica, 54
Germ cells
  germline determination
    conserved genetic machinery for, 51–52
    induction, 49
    maternal determination, 49
    primordial stem cells (PriSCs), 51
  morphological features, 48–49
Germline determination
  conserved genetic machinery for, 51–52
  induction, 49
  maternal determination, 49
  primordial stem cells (PriSCs), 51
Gland cells, 14
Glutamatergic neuron, 106
Glutamatergic synapse, 106
Gnathostomulida, 133
G-proteins, 106

H

Helobdella robusta, 144
Hexapoda, 132
Hydrozoans, 108

I

Integrins, 83–84
Intermediate mesoderm (IM), 141
Ionotropic receptors, 106

L

Leucosolenia complicata, 112
Lipophil cells, 14

M

Macroorganisms, 27
Malpighian systems, 148
Mesenchyme-to-epithelial transition (MET), 141
Mesoderm, 130
Mesonephros, 138
Metanephridia, 149
  aglomerular nephron, 138
  in amniotes, 139
  mesonephros, 138
  morphological diversity of, 137
  phoronids/annelids, 136
  ultrafiltration, 136
Metanephridial systems, 136–139
Metazoan epithelium
  adhesive junctions, 91
  ancestral basement membrane, 86–88
  communicating junctions, 89
  non-bilaterian relationships, 87–88
  sealing/occluding junctions, 89–90
Mnemiopsis leidyi, 110
Myriapoda, 132

N

Nematocysts
  bioinformatic mining, 40
  and dinoflagellates, 38
  immunolabeling, 41
  minicollagens, 40
  morphotypes, 34
  Polykrikos kofoidii, 40
  position and structure of cnidarian nematocysts, 35
  structure of, 37
  transcriptomics, 40
  in warnowiids, 40
Neofunctionalization, 5
Nephridia, 149
Nephrozoa, 131
Neuron
  choanocytes, 118
  epithelia, 105
  excretory cells, 105
  gene markers

Index

in *C. elegans*, 116
neuronal cell populations, 116
N. vectensis scRNAseq study, 116
placozoan *Trichoplax adhaerens*, 117
in *Pleurobrachia bachei*, 115
scRNAseq approach, 115
single cell RNA sequencing data, 116–117
transcription factors, 115–116
G-protein coupled receptors (GPCR), 104
interactions neurons, 105
label clusters, 118
marker genes, 118
metazoan phylogeny, 117
muscle cells, 104
neuronal phenotype, 104
neurotransmitters, 119
in non-bilaterians, 105–109
polarity/elongate processes, 119
scRNAseq methods, 118
single cell sequencing approach, 118

O

Ocelloids
cell division, 30
eyespots, 31
lens-bearing photosensory structures, 29
ocelloid ultrastructure, 31
in phototaxis, 29
of warnowiid dinoflagellates, 30

P

Paratomella rubra, 133
Patyhelminths, 105
Pax-Six-Eya-Dach network (PSEDN), 112
Peptidergic cells, 14
Pigoraptor vietnamica, 18
Pinacocytes, 14
Placozoans, 108
Platyhelminthes, 62–63
Pluripotent stem cells, 53
Polykrikos kofoidii, 40
Porifera, *see* Sponges
Post-synaptic neuron membrane, 106
Pre-synaptic neuron, 106
Primordial stem cells (PriSCs), 51, 56
Protonephridia, 133–136, 146, 149

R

Rhogocytes, 140

S

Salpingoeca rosetta, 18, 19
Scyphozoans, 109

Secretory excretory organs, 132
Sensory cilium, 111
Sensory-neuronal cell types
actin microfilaments, 109
Amphimedon queenslandica, 113
Beroe abyssicola, 109
in ctenophores, 109–111
deep sea glass sponges, 119
electrical recordings, 102
endopinacocytes, 114
eukaryotes, 102
hypotheses of origins, 103
invertebrate nervous systems, 103
in *Mnemiopsis leidyi*, 110
new fixation and staining techniques, 101
in placozoans, 112, 113
single cell sequencing (scRNAseq), 104
sponge larva, 112
vertebrate neuron, 102
Signaling proteins, 106
Single-cell genomics, 22
Spermatogenesis, 3
Sphaeroforma arctica, 18, 20
Sponge embryogenesis, 57–58
Sponges
archaeocytes and choanocytes
ancestral primordial germ cell, 56
asexually and sexually, 53
basophilic cytoplasm, 53
demosponges and hexactinellids, 55
in *Geodia atlantica*, 54
molecular and genetic techniques, 56
morphological traits, 53
origin of gametes, 53
pluripotent stem cells, 53
primordial germ cells, 56
sponge sperm, 53
determination of germline, 57–58
flagellum of choanocytes, 1
molecular signatures of germ cells, 56–57
morphological features of germ cells, 56
Statocyst
animal statocysts, 32
in *Euglena*, 32
loxodids, 33
and Muller's vesicle, 33
Trichoplax, 34
Subfunctionalization, 5
Sycon calcavaris, 112
Sycon ciliatum, 112

T

Tissue regeneration, 15
Transmembrane proteins, 129
Trichoplax, 34
Trichoplax adherens, 113

Tunicate nephrocytes, 140

U

Unicellular eukaryotes, 27
Unicellular holozoans
   Capsaspora owczarzaki, 19
   choanoflagellates, 16, 18
   colonial choanoflagellate, 19
   Corallochytrea, 18
   Corallochytrium limacisporum, 18
   DNA replication/RNA metabolism, 20
   filastereans, 16
   ichthyosporeans and corallochytreans, 18
   Pigoraptor chileana, 20
   Pigoraptor vietnamica, 20
   RNA-Seq analyses, 18, 19
   teretosporeans, 16, 20
   transcriptomic analysis, 21

V

Vesicular GABA transport proteins (VGATs), 106
Vesicular glutamate transport proteins (vGLUTs), 106
Volvocacea, 48

X

Xenacoelomorpha, 131

For Product Safety Concerns and Information please contact our EU representative GPSR@taylorandfrancis.com Taylor & Francis Verlag GmbH, Kaufingerstraße 24, 80331 München, Germany

Printed and bound by CPI Group (UK) Ltd, Croydon, CR0 4YY
08/06/2025
01896985-0009